# Perspectives on the His
# of Chemistry

**Series Editor**

Seth C. Rasmussen, Department of Chemistry and Biochemistry, North Dakota State University, Fargo, ND, USA

Commonly described as the "central science", chemistry and the chemical arts have an extremely long history that is deeply intertwined with a wide variety of other historical subjects. Perspectives on the History of Chemistry is a book series that presents historical subjects covering all aspects of chemistry, alchemy, and chemical technology.

Potential topics might include:

- An updated account or review of an important historical topic of broad interest
- Biographies of prominent scientists, alchemists, or chemical practitioners
- Translations and/or analysis of foundational works in the development of chemical thought

The series aims to provide volumes that advance the historical knowledge of chemistry and its practice, while also remaining accessible to both scientists and formal historians of science. Volumes should thus be of broad interest to the greater chemical community, while still retaining a high level of historical scholarship. All titles should be presented with the aim of reaching a wide audience consisting of scientists, chemists, chemist-historians, and science historians.

All titles in the book series will be peer reviewed. Titles will be published as both printed books and as eBooks. Both solicited and unsolicited manuscripts are considered for publication in this series.

More information about this series at https://link.springer.com/bookseries/16421

Marelene Rayner-Canham ·
Geoff Rayner-Canham

# Pioneers of the London School of Medicine for Women (1874–1947)

## Their Contributions and Interwoven Lives

Marelene Rayner-Canham
Department of Physics (retired)
Grenfell Campus, Memorial University
of Newfoundland
Corner Brook, NL, Canada

Geoff Rayner-Canham
Department of Chemistry
Grenfell Campus, Memorial University
of Newfoundland
Corner Brook, NL, Canada

ISSN 2662-4591 ISSN 2662-4605 (electronic)
Perspectives on the History of Chemistry
ISBN 978-3-030-95441-3 ISBN 978-3-030-95439-0 (eBook)
https://doi.org/10.1007/978-3-030-95439-0



This Springer imprint is published by the registered company Springer Nature Switzerland AG
The registered company address is: Gewerbestrasse 11, 6330 Cham, Switzerland

# Preface

In our research on pioneering British women chemists [1], we were surprised to discover a 'haven' for women chemists at the London School of Medicine for Women (LSMW). Until our own initial report [2], nothing had ever been published on their lives and contributions. This book was initially designed to build upon our preliminary study.

However, it was pointed out to us that it was important to frame the women chemists within the context of the LSMW. The LSMW had, itself, only existed from 1874 to 1947, and few people today are aware that there was once in London a School of Medicine solely devoted to training women doctors. As a result, we were urged to provide this wider context.

Commencing our expanded research project, we realized that there was a greater and more significant story which needed to be told. There had been a partial account of the LSMW within a history of the Royal Free Hospital, which subsequently incorporated the LSMW [3]. There have also been individual biographies of three of the major figures involved in the founding of the LSMW: Sophia Jex-Blake [4], Elizabeth Garrett Anderson [5], and Edith Pechey [6].

However, up until the research for this book, there has not been a holistic account which covers the struggles for women to obtain a medical education (the reason for the formation of the LSMW) and the interplay among the individuals involved. Each of the women, Jex-Blake, Garrett Anderson, Pechey, and also Isabel Thorne, played unique roles. Without any one of them, the LSMW would not have happened or survived. Nor has there been any recognition of the contributions of the other pioneering women who participated in the endeavour.

Though it was women who were responsible for founding the LSMW, throughout its history, the medical professors were almost exclusively male. Thus the saga of the LSMW chemistry department, women-run for almost the entire existence of the School, provides a unique story in itself, and even more so within the framework of institution.

Such an account also needs an ending. In this case, an unhappy one. With the belief that women-only institutions were no longer needed, the LSMW was totally absorbed into the Royal Free Hospital. Subsequent to the event, our research of issues of the *Royal Free Hospital Journal* shows it no longer to be a beacon for women, but one where the very idea of women doctors became questioned—history indeed repeating itself.

Corner Brook, Canada

Marelene Rayner-Canham
Geoff Rayner-Canham

## References


1. Rayner-Canham, M., & Rayner-Canham, G. (2020). *Pioneering British women chemists: Their lives and contributions*. London: World Scientific Publishing.
2. Rayner-Canham, M., & Rayner-Canham, G. (2017). Women chemists of the London School of Medicine for Women, 1874–1947. *Bulletin for the History of Chemistry*, *42*, 126–132.
3. McIntyre, N. (2014). *How British women became doctors: The story of the Royal Free Hospital and its Medical School*. London: Wenrowave Press.
4. Roberts, S. (1993). *Sophia Jex-Blake: A woman pioneer in nineteenth-century medical reform* (1st. ed.). London: Routledge.
5. Manton, J. (1965). *Elizabeth Garrett Anderson: England's first woman physician*. London: Methuen.
6. Lutzker, E. (1973). *Edith Pechey-Phipson, M.D.: The story of England's foremost pioneering woman doctor*. New York: Exposition Press.

# About This Book

This is a unique account of pioneering women and their interwoven lives which led to the founding of the London School of Medicine for Women (LSMW) in 1874. For the subsequent decades, this saga morphs into an account of the women-run chemistry department of the LSMW. At the end of our journey, we revert to the wider perspective of the demise of the LSMW in 1947, especially the largely misogynistic atmosphere that prevailed once it had been fully absorbed into the Royal Free Hospital.

## Chapter 1: Women as Apothecaries

There had been no direct path of admittance of women into the medical profession. Elizabeth Garrett found a devious route to a medical qualification through the apothecaries' examinations. However, this route was quickly blocked.

## Chapter 2: Women as Pharmacists

Pharmacy was seen as another preparatory route into medicine by women. In this chapter, the fight for the admission of women into the pharmacy profession is outlined.

## Chapter 3: Sophia Jex-Blake and Elizabeth Garrett (Anderson)

This chapter provides a summary of the lives and work of the two key individuals in the story of the formation of the LSMW. With totally different personalities, their lives repeatedly interacted, making it appropriate to combine their life-stories in the same chapter.

## Chapter 4: The Crucial Role of the 'Edinburgh Seven'

With the apothecary and pharmacy routes to medicine ceasing to be available to women, Sophia Jex-Blake applied for admission to the medical School of the University of Edinburgh. Told that they could not provide facilities for just one woman, Jex-Blake recruited six others, and more subsequently, to join the classes at the University. Unfortunately, their hopes were dashed.

## Chapter 5: Edith Pechey

A key figure, especially in the Edinburgh context, was Edith Pechey. In particular, she was at the centre of the cause célèbre of the award for excellence in chemistry. Later, she became a member of the first class of the LSMW and continued to be active in its support.

## Chapter 6: Others of the 'Edinburgh Seven'

Though the three women discussed so far were central to the founding of the LSMW, the other five also played a role and their lives must be included for completeness.

## Chapter 7: Women as Lady Doctors

A career in medicine was seen as one of the few paths of employment encouraged for middle-class Victorian girls. The printed media, both books and magazine articles, emphasized the need for lady doctors.

## Chapter 8: The Founding and Early Years of the LSMW

With the Edinburgh route closed, Jex-Blake and Garrett, together with Isabel Thorne, decided the only path to a medical qualification was the formation of a medical School specifically for women. In this chapter, we describe the challenges the founders faced in staffing the School, then in finding an associated hospital, and finally having the qualification accepted.

## Chapter 9: Pioneer Women of the LSMW

The School would not have been possible without the first eager cohorts of students. Nearly all of them had come with Jex-Blake from Edinburgh, while others had come from diverse backgrounds. In this chapter, we look at the lives of most of these women.

## Chapter 10: Chemistry at the LSMW

The staffing of the LSMW was almost exclusively male medical personnel throughout its existence. However, there was an exception: the chemistry department. This was almost entirely women-run throughout the 73 years. In this introductory chapter on the department, using quotes from the *LSMW Magazine*, we gain an insight into the lives and thoughts of the women students.

## Chapter 11: Lucy Everest Boole

Boole, who had started her professional life with pharmacy training, was the first woman Lecturer in chemistry. Her career was cut short, almost certainly, by her research involving an oil containing many toxic compounds, resulting in her death.

## Chapter 12: Clare de Brereton Evans

The successor to Boole was de Brereton Evans. Though she was the author of an influential book chapter on the teaching of chemistry to girls, she decided that research was her preferred path leading to her resignation.

## Chapter 13: Sibyl Taite Widdows

Widdows was to be the stalwart of the chemistry department for over 40 years. During that period, she was the subject of several satirical poems and rhymes in the *LSMW Magazine* which we incorporate in this chapter.

## Chapter 14: Phyllis Sanderson and Anne Ratcliffe

Following Widdows, Sanderson and Ratcliffe taught through into the post-War era, and even surviving the total absorption of the LSMW into the Royal Free Hospital (RFH) in 1947.

## Chapter 15: Other Chemistry Staff

In the preceding four chapters, the lives and contributions of the Heads of the chemistry department were described. There were other staff members in the chemistry department, several of whom made significant contributions to its running, and they are included in this chapter.

## Chapter 16: The End of the LSMW

Though this chapter is titled the end of the LSMW, in fact, its focus is more the aftermath of the absorption into the RFH. The ethos and women-supportive environment vanished, to be replaced by, at least in part, a very misogynistic tone of the male institution into which the LSMW had disappeared.

# Contents

# About the Authors

**Marelene Rayner-Canham** is a Physics Instructor (retired) at the Memorial University, Canada, and together with Geoff Rayner-Canham, has been contributing over the past 30 years to the history of women in science and chemical education with several published books, chapters and refereed articles about forgotten and pioneering women chemists. She was also involved in the organisation of several symposia on the history of chemistry and women in chemistry.

**Geoff Rayner-Canham** is an Honorary Research Professor of Chemistry at the Memorial University, Canada, and in the past 30 years he has been contributing to the history of women in science and the history of chemistry with several books, chapters and refereed articles published in this topic. Professor Rayner-Canham received several awards during his career, including the 3M Teaching Fellowship Award (2007), the Chemical Education Award, Chemical Institute of Canada (2008), and the NSERC Award for Science Promotion (2010), and in 2017 he was elected Fellow of the Royal Society of Chemistry. Together with Marelene Rayner-Canham, he organized several symposia on the history of chemistry and women in chemistry.

# Abbreviations

| | |
|---|---|
| BMA | British Medical Association |
| BMHW | Birmingham and Midland Hospital for Women |
| CLC | Cheltenham Ladies' College |
| EHDWC | Edinburgh Hospital and Dispensary for Women and Children |
| ELEA | Edinburgh Ladies Educational Association |
| ESMW | Edinburgh School of Medicine for Women |
| FMS | Female Medical Society |
| GMCGBI | General Medical Council of Great Britain and Ireland |
| KQCPI | King and Queen's College of Physicians in Ireland |
| LRFHSMW | London Royal Free Hospital School of Medicine for Women |
| LSMW | London School of Medicine for Women |
| NEHWC | New England Hospital for Women and Children |
| NHW | New Hospital for Women (London) |
| RFH | Royal Free Hospital |
| RFHSM | Royal Free Hospital School of Medicine |
| RHC | Royal Holloway College |
| UCL | University College, London |

# Women as Apothecaries

# 1

*To provide a complete framework for this monograph, it is important to introduce the avenues available for women to enter health-related professions. There were three such routes, the first to open up was that of Apothecaries and Apothecaries' Assistant (the topic of this chapter); pharmacist, the topic of* Chap. 2*; and then the fight for admission to medical School covered in later chapters. Out of necessity, women rarely followed a simple career path. Many tried whatever combination of paths enabled them to obtain their goal of a career in medicine: including the route of Apothecary.*

## Society of Apothecaries

Apothecaries had been involved in the production and dispensing of patent medicines from the Middle Ages [1]. There are records of women being registered as Apothecary or 'chymist and druggist' as far back as the 1600s [2]. Apothecaries were a recognized Guild with its own Court of Examiners to licence those who wished to dispense the herbal remedies of the time [3]. Their first encounter with a woman candidate, Elizabeth Garrett, occurred in 1865 [4].

Elizabeth Garrett (Fig. 1.1) was a key individual in the story of the London School of Medicine for Women (LSMW), and her life will be discussed more fully in later chapters, especially Chap. 3. Her long-held desire was to qualify as an M.D. and have her name added to the Medical Register. However, Garrett had been refused by every medical School to which she had applied on the grounds of her gender [5]. She then focussed her attention on becoming an Apothecary, as, at the time, passing the Apothecaries' Examination entitled its holder to a coveted place on the Medical Register and enabled the certified Apothecary to practice as a medical doctor (without a formal medical degree) [6].

Having taken private lessons with an Apothecary, Garrett enquired about the possibility of being examined at Apothecaries' Hall. This request caused great



M. Rayner-Canham and G. Rayner-Canham, *Pioneers of the London School of Medicine for Women (1874–1947)*, Perspectives on the History of Chemistry,
https://doi.org/10.1007/978-3-030-95439-0_1

**Fig. 1.1** Elizabeth Garrett. Public domain: https://en.wikipedia.org/wiki/Elizabeth_Garrett_Anderson#/media/File:Elizabeth_Garrett_Anderson.jpg; Photograph by Walery, published by Sampson Low & Co. in February 1889

consternation among the members of the Court of Examiners. Their legal counsel advised them that the Apothecaries' Act of 1815 opened the Apothecary examinations to all persons and as, by British law, women were persons, they could not be excluded. Garrett passed the Apothecaries' Preliminary Examination and then both parts of the Professional Examination, obtaining her diploma in 1865.

The Council realized that Garrett's success would encourage other women to follow suite. One of these was Isabel Thorne (see Chap. 6). Thorne, too, was endeavouring to obtain a medical qualification. Thorne recalled [7]:

> Miss Garrett had shown that the Society of Apothecaries was not quite impregnable, so with my brother-in-law's support I applied to the Society of Apothecaries for leave to present myself at the examination in Arts for admission to the L.S.A., and at last received permission on condition that if I satisfied the examiners it would give me no right to be a candidate at the subsequent medical examinations as it would have done in the case of a man. I was the only woman up for the examination in April 1868, and got through, but all further progress was barred as there was no medical school in Great Britain to which women were admitted.

The portents of more women in the pipeline persuaded the Council to act. Before admission to the Apothecaries' Examination, the candidate had to provide a certificate of attendance at the required lectures. As women were excluded from the lectures at public Schools, they had to be tutored privately. To block this avenue, the Court of Examiners announced that they would cease to accept the validity of certificates of attendance at private lectures. This tactic proved an effective means of circumventing the law and blocking the aspirations of subsequent women candidates.

## Apothecaries' Assistants

In addition to the profession of Apothecary, the Society controlled the examination leading to the lesser-rank of Apothecaries' Assistant. Women continued to have access to the examinations for this qualification. In the 1800s, Apothecaries' Assistants were required by any general practitioner or dispensary who dispensed their own medicine. It was in 1887 that Fanny Saward had become the first woman to gain admission to the Apothecaries' Assistants' Examination.

The Assistants' certificate was, as Penelope Hunting remarked [8]: "... a post suited to a young lady willing to work as the handmaiden of the doctor for little pecuniary reward." This occupation became a female ghetto [9]; for example, in 1917, nine men and 233 women registered for the Apothecaries' Assistants' Examination. Among those women was an Agatha Mary Clarissa Christie, who later gained fame as a writer of detective stories. However, some women did use Apothecaries' qualifications as a stepping-stone towards a medical degree, as we will show in later chapters.

## Commentary

As the pharmaceutical industry developed during the 1930s, greater and greater scientific knowledge was demanded of the dispensers of medicines. For this reason, the more basic examinations of the Apothecaries' Hall fell out of favour compared with the more lengthy and rigorous studies demanded of those holding qualifications of the Pharmaceutical Society. The Pharmaceutical Society (the subject of the next chapter) was seen by women as another avenue by which they might climb the ladder of medical education, and ultimately gain a medical degree.

## References

1. Hunting, P. (1998). *A history of the society of apothecaries*. The Society of Apothecaries.
2. Rawlings, F. H. (1984). Two 17th century women apothecaries. *Pharmaceutical Historian, 14*(3), 7.
3. Woolf, J. S. (2009). Women's business: 17th century female pharmacists. *Chemical Heritage, 27*(3), 20–25.
4. Ref. 1, Hunting, pp. 207–210.
5. Manton, J. (1965). *Elizabeth Garrett Anderson*. Methuen.
6. Rayner-Canham, M., & Rayner-Canham, G. (2008). *Chemistry was their life: Pioneer British women chemists, 1880–1949* (pp. 383–385). Imperial College Press.
7. Thorne, I. (1905). *Sketch of the foundation and development of the London School of Medicine for Women* (p. 5). G. Sharrow.
8. Ref. 1, Hunting, p. 230.
9. Adams, D. W. (2010). *The rise and fall of the Apothecaries' Assistants 1815–1923*. Ph.D. Thesis, U. Hertfordshire. https://core.ac.uk/download/pdf/1641247.pdf. Accessed 30 Sept. 2021.

# 2 Women as Pharmacists

*The avenue of entering medical studies by means of an Apothecary qualification (see* Chap. 1) *had been closed. However, for middle-class young women, a pharmacy qualification was seen as an alternative stepping-stone into medicine, should medical Schools subsequently open their doors to women.*

## Careers in Pharmacy

In 1868, *The Englishwomen's Review* had pronounced pharmacy as a suitable profession for a woman [1]. By the 1880s, the need for careers for young women became a topic of concern. Published in 1884, Arthur Talbot Vanderbilt authored a book titled: *What to do with Our Girls*. In a chapter: 'Lady Chemists and Druggists,' he proposed that the career of pharmacy was particularly suited to [2]: "… daughters of country medical men, who in many instances have acquired some practical knowledge of dispensing, and of the properties of the various drugs."

This belief that pharmacy was an ideal career for daughters of medical doctors was reiterated over two decades later. Professor J. E. Walden of the Westminster College of Chemistry and Pharmacy wrote an article on pharmacy as a career for women in the *Girl's Realm Annual* of 1907–1908. The article was illustrated with two photos of groups of 'girl chemists' in the laboratory. As to becoming a 'girl chemist,' in view of the 'severity' of the Pharmaceutical Society Examination, Walden suggested that girls consider the Apothecaries' Examination as a stepping-stone. He commented that [3]:

> The girls who take up the profession are principally doctors' daughters, or other relatives, those who have some means, and yet want something to do. A great many nurses also take it up, for the dispensing certificate is a valuable asset to the trained nurse. It is, however, generally suitable for the well-educated of the middle classes.



M. Rayner-Canham and G. Rayner-Canham, *Pioneers of the London School of Medicine for Women (1874–1947)*, Perspectives on the History of Chemistry,
https://doi.org/10.1007/978-3-030-95439-0_2

A second argument for pharmacy as a career pathway for young women was the need for women pharmacists to work with women doctors in developing countries. In 1887, pharmacist Isabella Skinner Clarke-Keer authored an article in the girls' magazine *Atalanta* which promoted pharmacy as a career path for girls [4]. Clarke-Keer saw particularly great opportunities for women pharmacists in India [4]:

> One great opening for women Pharmacists is in India, where so many of the women doctors are now settling. Women Pharmacists should accompany them, for in many instances in consequence of the present want of women dispensers, the doctors are obliged to dispense their own medicines. ... the work should be undertaken by the trained and qualified Pharmacist ...

## Pharmacy Examinations

The Pharmaceutical Society, founded in 1841, was given the authority to organize the required professional examinations [5]. To become a licenced pharmacist, there were three separate examinations: the Preliminary, the Minor, and the Major.

The Pharmaceutical Society Preliminary Examination was more of a skills test in Latin, French, and arithmetic, and it entitled the successful candidate to be registered as an Apprentice. Passing the Pharmaceutical Society Minor Examination resulted in the designation as an Assistant to a chemist or druggist, while passing the Major allowed the graduate to call themselves a Pharmaceutical Chemist. The Pharmaceutical Society Major Examination was described as 'decidedly difficult', and it focussed on advanced chemistry, *materia medica*, and botany. For Pharmacy Assistants who had actively been involved in the profession for three years, passing the Modified Major examination was all that was necessary [6].

By the early 1800s, women pharmacists were to be found all over the country [7]. In fact, when the first compulsory Register of all practising pharmacists was undertaken in 1869, 215 of the 11,638 registered chemists and druggists were women. Most of the women were continuing a business that had been started by a father or husband who had subsequently died.

The first woman to take the examinations required under the 1868 Act was an F. E. Potter, who applied in 1869 to take the Modified Major Examination. It was only when the individual appeared to sit the examination was it realized she was a woman—Frances (Fanny) Elizabeth Potter. As the Act made it clear that the Society had a duty to examine all persons, Potter was allowed to take the exam, which she successfully passed. Potter (Mrs. Deacon) was followed six months later by Catherine Hodgson Fisher [7].

In 1873, Alice Vickery became the first woman to pass the Minor. However, like some others, she was only studying pharmacy with an intent to use it as springboard into a medical education. That same year, Vickery travelled to France, being admitted to study medicine at the University of Paris. She returned to London in 1877, registering as a student at the London School of Medicine for Women (LSMW). At the LSMW, Vickery accomplished her goal of an M.D. in 1880 [8].

## The School of Pharmacy

The Pharmaceutical Society had its own School of Pharmacy. In 1861, the discovery that a 'lady' had gained admission to the lectures of the School caused a flurry of concern. As she was already in attendance, the Library, Museum and Laboratory Committee of the Society had little choice in the matter, but they added that ladies [9]: "must be regarded as attending upon sufferance".

The 'lady' seeking admission had been Elizabeth Garrett (see Chap. 1), who had wished to use the lectures of the School of Pharmacy to prepare herself for the Apothecaries' Examination. A student at the time, Michael Carteighe (later President of the Pharmaceutical Society), recalled [10]:

> I remember about the time I became a student there that we had a lady student in this house. ... I remember the envious eyes with which a number of us regarded her. I do not think we regarded her with envy because she was a lady—in fact, we admired her on that account: but we were conscious that when once a lady comes into a class she means to take prizes, and I am afraid we were selfish enough to think of that rather than anything else.

## Women's Admission to the Pharmaceutical Society

Though entry to the profession had been comparatively easy, admission to the professional body, the Pharmaceutical Society, proved challenging [11]. Even though, by law, the Society was forced to admit women to its examinations, the Society acted on the premise that pharmacy was a male profession and that the Society itself was a male preserve. In fact, the assumption had been that the practice of female relatives taking over a pharmacy would cease once Registration became law. Thus, they were unprepared for the 'women problem' to be a regular business item for the Council of the Pharmaceutical Society for the next decade.

The first woman to apply for membership of the Pharmaceutical Society was Elizabeth Leech in 1869 [12]. Leech had learned her pharmacy skills from her father, having worked in his shop for seven years. Following his death, for six years she shared the running of the shop with her brother and then on her own for another nine years. However, the Lancashire cotton famine had forced her out of business. Her application noted that she believed that membership in the Pharmaceutical Society would help her resume her business. The Council rejected her request a total of three times, the last being in 1872.

It is often overlooked how much the progress of women relied on sympathetic men [13], and the cause of women in pharmacy was no exception. The 'white knight' in this case was Robert Hampson [6]. In 1872, Hampson was elected to the Council of the Pharmaceutical Society, focussing his activist radicalism on the issue of women's rights. Annie Neve recalled [14]: "Women pharmacists have good reason to honour the memory of Robert Hampson ... As a member of the [Pharmaceutical] Society's Council he repeatedly pleaded and wrought for the admission of women pharmacists to membership of the Society."

At the Council meeting of 5 February 1873, there were names of three women put forward among the 166 candidates for admission as 'registered students' of the Society [15]. These were Rose Minshull, Louisa Stammwitz, and Alice Rowland (Mrs. Hart). Minshull had attained the highest mark in the Preliminary examination, while Stammwitz was about mid-way in the list. Rowland had a Certificate from the Society of Apothecaries *in lieu* of taking the examination.

Hampson moved their election to the Society, and the President of the Society, George W. Sandford promptly moved an amendment that they [16]: "... be not elected ...". Heated debate followed. The final outcome of events was that women were barred from admission. Sandford articulated his position in the Correspondence pages of the *Pharmaceutical Journal* [17]: "I have always held that the Pharmaceutical Society was intended to be a Society of men, that certain disadvantages would arise from its being a mixed Society of men and women, ...".

Alice Rowland had only intended to obtain a pharmaceutical qualification as a potential pathway towards a medical degree. This route having failed, she subsequently attended the LSMW when it opened (see Chap. 8), then completed a medical degree at the University of Paris. Rowland married Ernest Hart, a medical doctor who strongly supported women in medicine. She subsequently wrote a highly acclaimed book: *Diet in Sickness and in Health* [18].

However, the issue of the admission of women would not go away. The names of Rose Minshull and Isabella Clarke-Keer were put forward again at the Council meeting of 1 October 1879, as *The Chemist and Druggist* reported [19]: "Mr. Hampson moved that they should be elected. He ... urged that it was the duty of the Council to elect all eligible persons irrespective of sex." The outcome this time was very different. When the vote was called, only Sandford, still the President of the Society, was against their admission. The opposition to women members had collapsed; this particular battle had been won. Louisa Stammwitz applied and was accepted a year later.

## Commentary

Becoming an Apothecary or Pharmacist were two routes by which had been possible to 'back-door' into a medical qualification. Garrett had effectively closed both avenues for subsequent women who were seeking a route into medical education. The only option now was to find a direct route into a medical School. It was to be Garrett's friend, Sophia Jex-Blake, who would make a direct assault upon the medical School of the University of Edinburgh, as we will show in Chap. 4.

## References

1. Bernard, B. (1868). Pharmacy as an employment for women. *The Englishwomen's Review (First Series), 1*, 348.
2. Vanderbilt, A. T. (1884). *What to do with our girls* (p. 88). Houlston & Sons.

3. Walden, J.E. (1907–1908). Girls as chemists. How a girl may take up the work of chemistry, with a view to keeping a pharmacy, or becoming a doctor's dispenser. *Girl's Realm Annual, 10*, 396.
4. Clarke-Keer, I. S. (6 Oct. 1887). Employment for girls. Pharmacy. *Atalanta*, 295.
5. Holloway, S. W. F. (1991). *Royal pharmaceutical society of great Britain 1841–1991: A political and social history*. The Pharmaceutical Press.
6. Jordan, E. (1998). 'The great principle of English fair-play': Male champions, the English women's movement and the admission of women to the Pharmaceutical Society in 1879. *Women's History Review, 7*(3), 381–409.
7. Burnby, J. G. L. (1990). Women in pharmacy. *Pharmaceutical the Historian, 20*(2), 6.
8. Anon. (13 Oct. 2021). Alice Vickery. https://en.wikipedia.org/wiki/Alice_Vickery. Accessed 13 Oct. 2021.
9. Ref. 5, Holloway, p. 262.
10. Cited in: Hudson, B. (2013). *The school of pharmacy university of London: Medicines, science and society, 1842–2012* (p. 63). Academic Press.
11. Rayner-Canham, M., & Rayner-Canham, G. (2008). *Chemistry was their life: Pioneer British women chemists, 1880–1949* (pp. 385–390). Imperial College Press.
12. Jordan, E. (2001). Admitting a dozen women into the society: The first women members of the British Pharmaceutical Society.
13. Strauss, S. (1982). *Traitors to the masculine cause: The men's campaigns for women's rights*. Greenwood Press.
14. Neve, A. (1926). Miss Annie Neve's reminiscences of Mrs. Clarke-Keer. *Pharmaceutical Journal and Pharmacist, 117*, 375.
15. Anon. (1872–1873). Transactions of the Pharmaceutical Society: Meetings of the Council, February 5th, 1873. *Pharmaceutical Journal and Transactions 3*, 629–631.
16. Ref. 15, Anon., p. 631.
17. Sandford, G. W. (1873). Correspondence. *Pharmaceutical Journal and Transactions, 3*, 698.
18. Hart, Mrs. [Rowlands, A.M.]. (1895). *Diet in sickness and in health*. The Scientific Press.
19. Anon. (1879). The pharmaceutical council: Ladies admitted as members. *Chemist and Druggist, 21*, 422.

# 3 Sophia Jex-Blake and Elizabeth Garrett (Anderson)

*Sophia Jex-Blake and Elizabeth Garrett (married name: Garrett Anderson) were the two women who did more than any others to advance the cause of the teaching of medicine to women. In particular, each of them played key roles in the founding of the London School of Medicine for Women. Though contemporaries with interwoven lives, in temperament they were extreme opposites: Jex-Blake, an ebullient extrovert, and Garrett, a calm introvert. Several books have been written about each of these individuals. Here, we will focus on the key factors in their lives which relate to the founding of the School.*

## Sophia Jex-Blake

Sophia Jex-Blake (Fig. 3.1) was born in Hastings, Sussex, on 21 January 1840, daughter of retired lawyer Thomas Jex-Blake and Mary Cubitt [1]. Until the age of eight, she was home-educated and then attended various private Schools in southern England. Asked to leave from some of the Schools, Jex-Blake was considered by her teachers as both brilliant and unmanageable.

Despite her parents' objections, in 1858, Jex-Blake enrolled at Queen's College, Harley Street. The following year, while still a student, she accepted an offer to become Mathematics Tutor at the College. Jex-Blake worked without pay: her family did not expect their daughter to earn a living, and indeed her father refused her permission to accept a salary.

Part of the time while she was at Queen's College, Jex-Blake lived with the family of Octavia Hill. Octavia Hill was Jex-Blake's first love [2]. One year older than Jex-Blake, Hill had tutored her in business studies. However, Jex-Blake's fiery temperament proved incompatible with Hill's calm and placid nature, and Hill severed the relationship in 1861. As a result of the rejection, Jex-Blake left Queen's College at the end of that term, returning to her parents' home with severe depression.



M. Rayner-Canham and G. Rayner-Canham, *Pioneers of the London School of Medicine for Women (1874–1947)*, Perspectives on the History of Chemistry,
https://doi.org/10.1007/978-3-030-95439-0_3

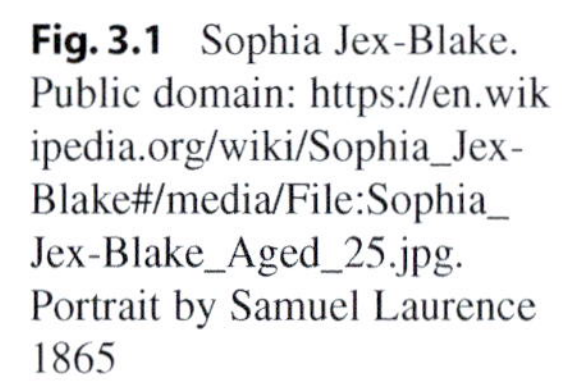

**Fig. 3.1** Sophia Jex-Blake. Public domain: https://en.wikipedia.org/wiki/Sophia_Jex-Blake#/media/File:Sophia_Jex-Blake_Aged_25.jpg. Portrait by Samuel Laurence 1865

To gain more teaching experience, Jex-Blake looked for a position in Germany. The German girls' Schools provided a more rigorous and academic education, and this was an attraction for her. In 1862, she obtained a temporary teaching appointment at the Grand Ducal Institute in Mannheim, Germany [3]. Jex-Blake found her first experience of teaching a trying one. Her biographer, Margaret Todd, who later summarized Jex-Blake's diary entries, wrote [4]:

> When the novelty wore off, the girls, after the fashion of their kind, began to try how far they could go with the English governess. ... S. J.-B. though an admirable teacher, did not show herself particularly strong in the matter of keeping order. The pupils found out their power of "tormenting" her, ...

Jex-Blake was intrigued by the co-educational Schools in the USA. In 1865, with her friend Isabel Bain, she travelled across the Atlantic to Boston, Massachusetts. Jex-Blake visited a range of educational establishments, compiling extensive studies which were published as *A Visit to Some American Schools and Colleges* [5]. At the New England Hospital for Women and Children (NEHWC) in Boston [6], she met one of the US pioneer woman physicians, Lucy Ellen Sewall, who became an important and lifelong friend. Before Jex-Blake had left Germany, she had contracted scarlet fever and a recurrence occurred in Boston. Sewall arranged for her care. Jex-Blake repaid Sewall's kindness by acting as accountant for the Hospital, a task for which she had an especial aptitude. Becoming absorbed into the medical activities, she took over the dispensary duties when there was a staff shortage. In addition, Jex-Blake also acted as hospital chaplain for a period, though eventually she decided that she lacked the certainty in her belief to become a minister.

Medicine, however, continued to fascinate Jex-Blake. She formally enrolled as a medical student in the NEHWC, an experience which led her to decide to aim for a medical degree at Harvard University. In 1867, Jex-Blake together with Susan

Dimock, a trainee at the NEHWC, wrote directly to the President and Fellows of Harvard University requesting admission to the University's Medical School. They received their reply a month later, in a letter which stated that there was no provision for the education of women in any department of this University.

Jex-Blake then moved to New York, where in 1857, Elizabeth Blackwell and Emily Blackwell, British-born American pioneer medical doctors, had established the New York Infirmary for Indigent Women and Children [7]. Then in 1868, the Blackwells established a medical College for women next to the infirmary. Named The Woman's Medical College of the New York Infirmary, it incorporated Elizabeth Blackwell's innovative ideas about medical education including much more extensive clinical training. Jex-Blake was one of the first students, but only a few months later, she received word that her father was dying. She took the first available ship across the Atlantic but arrived too late. He had died before Jex-Blake had even received the letter telling her he was ill.

It was in 1869 that Jex-Blake decided to attempt to gain admittance to the medical School of University of Edinburgh, a saga recounted in Chap. 4. After the attempt failed, she moved to London and, in 1874, was a key figure in the founding of the London School of Medicine for Women (LSMW). Jex-Blake continued to play an influential role with the LSMW until she resigned in 1896 and severed all connections with the School (see Chap. 8).

In January 1877, Jex-Blake passed the medical examinations of the University of Bern, finally being granted an M.D. Four months later, she qualified as a Licentiate of the King and Queen's College of Physicians in Dublin (see Chap. 9). This, at last, enabled her to be registered with the General Medical Council, the third registered woman doctor in Britain.

Jex-Blake then returned to Edinburgh, where, in June 1878, she set up her practice. Three months later, she opened an outpatient clinic where poor women could receive medical attention for a small fee. The clinic was moved to a larger site in 1885, and a small five-bed ward was added. This addition necessitated a name change to the Edinburgh Hospital and Dispensary for Women and Children (EHDWC), Scotland's first hospital for women, staffed entirely by women.

In 1886, Jex-Blake established the Edinburgh School of Medicine for Women (ESMW) [8]. The name suggested a structured professional institution similar to the LSMW. However, in fact, it was simply small informal classes taught by some pro-women male physicians with links to the University of Edinburgh. Jex-Blake's skill as a teacher did not match her accreditation as a doctor. An acrimonious dispute with her students resulted in an infamous court case in 1889, where Jex-Blake was successfully sued for damages. Dissatisfied with the quality of education from Jex-Blake at the ESMW, her former student, Elsie English, then set up a rival institution later in 1889: the Edinburgh College of Medicine for Women. To add to the competition, in 1892, the University of Edinburgh began accepting women students. With the bad press from the court case and the competition from these other institutions, the ESMW finally closed its doors in 1898. During its existence, the ESMW had educated approximately 80 women, 33 of them completing the program.

One of Jex-Blake's students at the ESMW had been Margaret Todd. It is likely that her relationship with Jex-Blake dated from this period. After graduating with an M.D. from the Free University of Brussels in 1894, Todd was appointed Assistant Physician at the EHDWC. When Jex-Blake retired in 1899, both she and Todd moved to Rotherfield in East Sussex. Jex-Blake died there in January 1912. Todd died in 1918, and in an Obituary for Todd in the *British Medical Journal*, it was noted that [9]: "The life of Miss Jex-Blake [authored by Todd]—an able and exhaustive piece of biography—made no mention of the affections and life work of her friend [Todd], and this omission is the more to be regretted now that death has closed the careers of both."

James Stansfeld, who had been closely associated with the London campaign (following the failure of the Edinburgh campaign), wrote a summary of the progress of women in medicine up to 1877. He concluded [10]:

> Dr. Sophia Jex-Blake has made the greatest of all contributions to the end attained. I do not say that she has been the ultimate cause of success. The ultimate cause has been simply this, that the time was at hand. It is one of the lessons of the history of progress that when the time for reform has come you cannot resist it, though if you make the attempt, what you may do is to widen its character or precipitate its advent. Opponents, when the time has come, are not merely dragged at the chariot wheels of progress—they help to turn them.

## Elizabeth Garrett

Elizabeth Garrett was born in London, on 9 June 1836, the second of eleven children of Newson Garrett and his wife, Louisa Dunnell [11]. When she was 10, a governess was employed to educate Garrett and her sister. Garrett despised her governess and sought to outwit the teacher in the classroom [12]. When Garrett was 13 and her sister 15, they were sent to a private School, the Boarding School for Ladies in Blackheath, London. Her main complaint about the School was the lack of science and mathematics instruction. After her basic schooling was complete, Garrett remained at home.

When Elizabeth Blackwell visited London in 1859, Garrett travelled there to attend Blackwell's lectures on *Medicine as a Profession for Ladies* [13]. By good fortune, Garrett was invited to a private post-lecture celebration. At the time, Garrett had not chosen a career path. However, Blackwell gained the impression that she was seriously considering a career in medicine and encouraged her to do so.

In 1860, Garrett decided to spend six months as a surgery nurse at the Middlesex Hospital, London. After proving her competence, she was allowed to attend an outpatients' clinic, then her first live operation. Garrett's attempt to enrol in the Medical School of the Middlesex Hospital was unsuccessful. However, she was allowed to attend private tuition in Latin, Greek, and *materia medica* with the Hospital's Apothecary, Joshua Plaskitt, while continuing her work as a nurse. Eventually, she was permitted into the dissecting room and to attend chemistry lectures.

Unfortunately, her outstanding intellect was to be her downfall. The male students who had previously tolerated her in class, expecting her presence to be fleeting, not permanent, submitted a petition to the Hospital administration that Garrett be expelled and that no further women students be permitted. The administration decided she had to go. Garrett wrote to her sister [14]: "I believe my death-warrant will be signed next Thursday." And after the 'death-warrant' was delivered to her, Garrett wrote to her father [15]: "You will be sorry to hear that the students have had their way. ... They don't like to see a woman working hard and they want to snuff her out if possible. Their masculine sense of superiority is insulted by the competition which must tacitly go on." Perhaps out of guilt, the Medical School provided her with an Honours Certificate in chemistry and *materia medica.*

Garrett then applied to several British medical Schools, all of which refused her admittance. There was one other route to a medical qualification: the Society of Apothecaries (see Chap. 1). By Royal Charter, the Society could examine and licence 'all persons' who completed five years as an apprentice and three years in lectures, demonstrations, and hospital practice. Plaskitt agreed to take her on as an apprentice in 1861.

In October of 1861, Garrett attended a series of lectures on physiology being given by Thomas Henry Huxley, biologist and anthropologist, who specialized in comparative anatomy. Expecting that she and her friend Ellen Drewry would be the only women in the audience, they were pleasantly surprised that there were three others: Sophia Jex-Blake and Octavia Hill, together with Octavia's sister, Miranda Hill. This was the first encounter of Garrett and Jex-Blake, not realizing their life-paths would become intertwined.

Garrett was admitted to the Apothecaries' Preliminary Examination in 1862. This left the requirement of the three years of lectures, demonstrations, and practice. Over the next three years, she studied privately with various professors, including some at the University of St. Andrews, Scotland; the Edinburgh Royal Maternity Hospital; and the London Hospital Medical School. In 1865, she finally took her exam and obtained a Licence (LSA) from the Society of Apothecaries to practise medicine. That year, only three of the seven candidates passed the exam, Garrett obtaining the highest marks. The Society of Apothecaries immediately amended its regulations to prevent other women obtaining a Licence (hence barring Jex-Blake from following the same route).

Though she was now a Licentiate of the Society of Apothecaries, as a woman, Garrett could not take up a medical post in any established hospital. With financial backing from her father, in 1866, she founded and became General Medical Attendant to St Mary's Dispensary for Women and Children, London.

Garrett had told Blackwell that her intent was to acquire an M.D. from a foreign University, as the Licence from the Apothecaries Hall had gained her little respect. She read that the Dean of the Faculty of Medicine at the University of Paris (aka the Sorbonne) was in favour of admitting women as medical students. Garrett applied for admission. Fortunately, Emperor Napoleon III was ill at the time, and the Council of Ministers was presided over by Empress Eugénie [16]. The Empress approved the admission of Garrett and two other women applicants before the ministers had an opportunity to discuss the issue.

In order to learn French and prepare for the four required examinations, Garrett suspended her social life. The second examination was particularly challenging as she had to perform two operations in front of the examiners and with an audience. Garrett also had to submit a thesis. Finally, in June 1870, the examiners announced her success. Garrett had become the first woman to earn an M.D. qualification from the University of Paris—Sorbonne.

Upon Garrett's return to England, she was interviewed for the position of Visiting Medical Officer for a new children's hospital in London. She was unsuccessful in obtaining the post. The vice-chair and financial officer of the hospital board was James Skelton Anderson, who worked for the Orient Steamship Company. He was impressed by Garrett's intellect and personality and arranged to meet her socially. A friendship developed, and they were married in 1871.

In 1872, Garrett transformed St Mary's Dispensary for Women and Children in London into the New Hospital for Women (NHW). This Institution enabled, for the first time, disadvantaged women to obtain medical help from qualified female practitioners.

Garrett's name was added to the Register of the British Medical Association (BMA) in 1873. However, as a result of her election, a motion was passed by the BMA to ban the further admission of women [17]. As Mary Ann Elston has pointed out, this was [11]: "... one of several instances where Garrett, uniquely, was able to enter a hitherto all male medical institution which subsequently moved formally to exclude any women who might seek to follow her."

The formation of the LSMW and its subsequent administration was to become the focus of the remainder of Garrett's life (see Chap. 8). She retired to Aldeburgh, Suffolk, in 1902, where she died in 1917.

## Commentary

It was in 1874 that Jex-Blake and Garrett came to the decision that the only way forward was to start an independent medical School for women in London, the subject of Chap. 8. However, as we will explore in the intervening chapters, the founding of the LSMW was not just the work of these two, but involved other key figures, particularly Edith Pechey and Isabel Thorne. These additional women became part of the campaign as a result of the failure to graduate from the University of Edinburgh, the saga described in Chap. 4.

## References

1. Roberts, S. (1993). *Sophia Jex-Blake: A woman pioneer in nineteenth-century medical reform* (1st ed.). Routledge.
2. Darley, G. (1990). *Octavia Hill: A life*. Constable.

3. Campbell, O. (2021). *Women in white coats: How the first women doctors changed the world of medicine* (pp. 90–92). Park Row Books.
4. Todd, M. (1918). *The life of Sophia Jex-Blake* (pp. 135–136). Macmillan and Co.
5. Jex-Blake, S. (1867). *A visit to some American schools and colleges*. Macmillan and Co.
6. Davis, A. T. (Apr. 1991). America's first school of nursing: The New England hospital for women and children. *Journal of Nursing Education, 30*(4), 158–161.
7. Nimura, J. P. (2021). *The Doctors Blackwell: How two pioneering sisters brought medicine to women and women to medicine*. W.W. Norton and Co.
8. Somerville, J. M. (2005). Dr. Sophia Jex-Blake and the Edinburgh School of medicine for women, 1886–1898. *Journal of the Royal College of Physicians, Edinburgh, 35*(3), 261–267.
9. Anon. (Sept. 1918). Scotland: The late Dr. Margaret Todd. *British Medical Journal, 2*, 299.
10. Stansfeld, J. (1877). Medical women. *Nineteenth Century, 1*, 901.
11. Elston, M. A. (1 Sept. 2017). Anderson, Elizabeth Garrett (1836–1917), physician. *Oxford Dictionary of National Biography*. https://doi.org/10.1093/ref.odnb/30406. Accessed 1 Oct. 2021.
12. Manton, J. (1965). *Elizabeth Garrett Anderson: England's first woman physician* (pp. 32–33). Methuen.
13. Ref. 3, Campbell, pp. 51–55.
14. Ref. 3, Campbell, p. 76.
15. Ref. 3, Campbell, p. 77.
16. Ref. 3, Campbell, pp. 204–209.
17. Lamont, T. (1992). The amazons within: Women in the BMA 100 years ago. *British Medical Journal, 305*, 1529–1532.

# 4 The Crucial Role of the 'Edinburgh Seven'

*The cause of the founding of the London School of Medicine for Women lay not in London, but in Edinburgh, at the University of Edinburgh. It was the thwarted medical ambitions of women, especially Jex-Blake, at that institution which led to the determination to found the School.*

## Edinburgh Ladies Educational Association

Before beginning the story, it is important to note that Sophia Jex-Blake and her colleagues had a powerful ally in Edinburgh: influential and affluent members of the Edinburgh Ladies Educational Association (ELEA). Founded in 1867 by campaigner Mary Crudelius, the ELEA responded to a demand for higher educational opportunities for middle-class Edinburgh women. Some members of the ELEA were subsequently to be instrumental in raising funds for the 'Edinburgh Seven'.

## The Beginning

In Chap. 3, we described how, following the death of her father in 1868, Jex-Blake returned to England. As Jex-Blake considered that Scotland had an enlightened attitude to education, she applied to study medicine at the University of Edinburgh in March 1869 [1]. The Medical Faculty and the *Senatus Academicus* both voted in favour of allowing her to study medicine. However, the University Court rejected her application on the grounds that the University could not make the necessary arrangements [2]: "in the interest of one lady".

The wording of this ruling, almost certainly unintentional, left the door open for a *group* of women to apply. Advertising in Scottish newspapers, Jex-Blake asked if any women wished to participate in a joint submission [2]. Of the respondents to the advertisement, she chose four initially, then added two more later. These six



M. Rayner-Canham and G. Rayner-Canham, *Pioneers of the London School of Medicine for Women (1874–1947)*, Perspectives on the History of Chemistry,
https://doi.org/10.1007/978-3-030-95439-0_4

individuals were: Mary Anderson, Emily Bovell, Matilda Chaplin, Helen Evans, Edith Pechey, and Isabel Thorne. Jex-Blake labelled them: '*Septem contra Edinam*' [3] (though as we discuss in Chap. 6, the numbers were actually more fluid). As Pechey had a unique role in the 'Edinburgh Seven' story, her life and career are separate in Chap. 5. The lives of the other applicants will be described in Chap. 6.

## Admission to the University of Edinburgh

Jex-Blake put the names forward to the University Court. The Court gave its approval, and in November 1869, these seven became the first women admitted to a British University. In the 1869 Calendar of the University, official regulations were inserted—reappearing annually for several years—that [4]:

> (1.) Women shall be admitted to the study of medicine in the University; (2.) The instruction of women for the profession of medicine shall be conducted in separate classes, confined entirely to women; (3.) The Professors of the Faculty of Medicine shall, for this purpose, be permitted to have separate classes for women; (4.) Women, not intending to study medicine professionally, may be admitted to such of these classes, or to such part of the course of instruction given in such classes, as the University Court may from time to time think fit and approve; (5.) The fee for the full course of instruction is such classes shall be four guineas; but in the event of the number of students proposing to attend any such class being too small to provide a reasonable renumeration at that rate, it shall be in the power of the Professor to make arrangements for a higher fee, subject to the usual sanction of the University Court; (6) All women attending such classes shall be subject to all the regulations now or at any future time in force in the University as to the matriculation of students, their attendance on classes, Examination, or otherwise; (7) the above regulations shall take effect as from the commencement of session 1869-70."

In a lecture given by Jex-Blake in 1872, she explained how the system worked for the two pre-med courses which they took; one being physiology and the other chemistry [5]:

> Though the lectures were delivered at different hours, the instruction given to us and to the male students was identical, and when the class examinations took place, we received and answered the same papers at the same hour and on identical conditions, having been told that marks would be awarded indifferently to "both sections of the class,"—this latter expression being, by the bye, repeatedly used during the course of the term by both the Professors who instructed us.

## Professor Crum Brown

The chemistry course was taught by Alexander Crum Brown, Professor and Chair of Chemistry [6]. To gain admission to Medical School, a student had to provide the authorities with a University Certificate of Attendance, to show that they had completed the prerequisite courses. Crum Brown refused to issue the Edinburgh Seven the Certificates of Attendance for the Chemistry Class. Instead, he offered them written certificates of them having attended [7]: a "ladies' class in the University." These, Jex-Blake derisorily referred to as Crum Brown's 'strawberry jam labels', as they were totally worthless in the context of admission to Medical School. Lacking the formal Certificates, the women were barred from the School. The women students appealed to the Senate of the University of Edinburgh. By a one-vote margin, the University Senate approved the issuing of University Certificates of Attendance to the women.

## The Surgeon's Hall Riot

Worse was yet to come. By July 1870, it had been agreed by the University that lecturers were free to lecture to mixed classes, if they wished [8]. In the Fall of 1870, the women students were members of a combined class in practical anatomy which was held at the Surgeons' Hall, outside of the University. All went well until the women arrived to take the anatomy exam in November.

As the women approached the Surgeons' Hall, they were mobbed by drunken male students. The Hall gates were slammed in their faces. Fortunately, one student, Tom Sanderson, who was already inside, saw their predicament, rushed out of the Hall, and managed to open the gates for them [9].

However, it was the aftermath of the exam which was more frightening for the women. Isabel Thorne recalled in the *Royal Free Hospital Magazine* [10]:

> By the end of the examination it was dark and a crowd had again gathered around the gates. We were asked if we would leave by a private door; but we felt it would not do to be intimidated, and relying on the support of our class mates, who formed a sort of bodyguard around us, arming themselves, in default of other weapons, with osteological specimens, we passed quietly through the mob, only our clothes being bespattered with the mud and rotten eggs thrown at us.

The attack on the women was disavowed the next day by many of the medical students. In a Letter to the Editor of *The Scotsman*, they cast the blame on chemistry students and, in particular, traced it back to their chemistry professor, Crum Brown [11]:

> Are only the hot-headed youths to be blamed who hustle and hoot at ladies in the public streets, and by physical force close the College gates before them? Or are we to trace their outrageous conduct to the influence of the class room, where their respected professor meanly takes advantage of his position as their teacher to elicit their mirth and applause, to arouse their jealousy and opposition, by directing unmanly innuendos at the lady students? ... The truth, however, is that the rioters were called together by a missive, circulated by the students in the ***Chemistry Class of the University*** [bold italic as in the original letter to the Editor] on Friday morning.

In the context of the continuing attacks on the women, it was a classmate by the name of Robert Wilson who came to their support. Following the riot, Wilson sent a letter to Edith Pechey to alert the women [12]:

> I wish to warn you that you are to be mobbed again on Monday. A regular conspiracy has been, I fear, set on foot for that purpose. ... I have made what I hope to be efficient arrangements for your protection. ...I had a meeting with my friend, Micky O'Halloran who is leader of a formidable band, known in college as 'The Irish Brigade' and he has consented to tell off a detachment of his set for duty on Monday. ... May I venture to hint my belief that the real cause of the riots is the way some of the professors run you down in their lectures. They never lose a chance of stirring up hatred against you. ... However, as I tell you, you and your friends need not fear, as far as Monday is concerned. You will be taken good care of.

In fact, the 'Irish Brigade' continued their escort duties of the women between accommodations and lectures for some time afterwards. Michael O'Halloran, an Irish male medical student at the time, has been overlooked as a hero of the event, having chosen to ally himself and his 'Brigade' with the women students, protecting them from what could have been severe assaults. At the time, it was possible for medical students to spend a year at different University medical Schools [13]. Sadly, we will never know what caused the noble gesture of O'Halloran and his Irish band in defending the women.

## The Final Insult

Despite several other attempts to impede their path, the women had passed all the examinations by 1872. However, the University of Edinburgh refused to grant them degrees. The group then took legal action against the University and, on 26 July 1872, initially won their case, the judges being scathing in their condemnation of the University [14].

The University appealed the ruling. Unfortunately, the appeal was upheld. The final verdict, given in 1873, stated that women should never have been admitted to the University in the first instance, and therefore could not graduate. Moreover, the women were compelled to pay all the legal costs, including the University's appeal, which amounted to the very significant sum at the time of £2,000. The *Committee to Secure a Complete Medical Education for Women in Edinburgh* came to their rescue, asking the public to provide financial aid and moral support, both of which were generously forthcoming. As the University of Edinburgh had refused to issue them degrees, the British Medical Association would not register the women as qualified doctors.

## A Plea to the University of St. Andrews

Sophia Jex-Blake did not give up. She submitted a petition on 17 July 1873 to the *Senatus Academicus* of the University of St. Andrews, signed by herself, Annie Reay Barker, Isa Foggo, Alice Ker, Sophy Jane Massingberd-Mundy, Agnes McLaren, Edith Pechey, Jane Russell Rorison, Elizabeth Vinson, and Elizabeth Walker (names which will appear again in later chapters). In this petition, Jex-Blake stated [15]: "The most general objection to the admission of women to the

Universities lies in the supposed difficulty of educating them jointly with male students of medicine." However, this argument did not apply to St. Andrews as it lacked a medical School. The petition informed the *Senatus Academicus* that: "... at least fifteen ladies would at once avail themselves of the permission, if given, to matriculate at the University of St. Andrews." The petitioners also offered to hire or build suitable facilities for a medical School and to arrange for lectures to be delivered in subjects not already covered in the curriculum at St. Andrews. The petition was rejected.

## A Plea to the House of Commons

Still Jex-Blake did not give up. In 1875, the Liberal M.P. William Cowper-Temple put forward a Bill entitling women to enter and graduate from Scottish Universities. Jex-Blake and 16 of her colleagues submitted a petition in support of the Bill, stating [16]:

> ... it has now been made illegal for any Scottish University to admit Women to instruction or graduation, and that your Petitioners are now thereby debarred from completing their Medical Education, and from obtaining Degrees from completing their Medical Education, and from obtaining Degrees from the University of Edinburgh, or from any other Scottish University, and are thus excluded from the Profession of Medicine, after the expenditure of much time, labour, and money with a view to their admission to the same.
> Your Petitioners therefore pray that your Honourable House will pass the Bill entitled "A Bill to remove doubts as to the powers of the Universities of Scotland to admit Women as Students, and to grant Degrees to Women."

The document was signed: Sophia Jex-Blake, Mary Edith Pechey, Isabel Thorne, Mary A. Marshall, Jane Russell Rorison, Matilda Chaplin Ayrton, Isabella Margaret Foggo, Elizabeth Ireland Walker, Alice J. S. Ker, Agnes McLaren, Anna Dahms, Emily Bovell, Sophy Massingberd-Mundy, Rose Anna Shedlock, Annie Reay Barker, Elizabeth Vinson.

At its Second Reading, the motion was rejected by 194 votes (mostly Conservatives) to 151 votes (mostly Liberals) [17].

## Commentary

Though these women had failed in their attempts for higher education in Edinburgh, and also at St. Andrews, from this setback, was to come a significant advance. Nearly all the women of the St. Andrews petition, and others, moved south to London and planned to form an independent medical School exclusively for women there. Thus, the founding of the London School of Medicine for Women (see Chap. 8) can be said to be the one fortunate outcome of the rejections.

## References


1. Jex-Blake, S. (1886). *Medical women: A thesis and a history* (2nd Ed.) (p. 71). Oliphant, Anderson, and Ferrier.
2. Ref. 1, Jex-Blake, p. 75.
3. Elston, M. A. (28 May 2015). Edinburgh Seven. *Oxford Dictionary of National Biography*. https://doi.org/10.1093/ref:odnb/61136. Accessed 21 Jan. 2020.
4. Ref. 1, Jex-Blake, p. 77–78.
5. Ref. 1, Jex-Blake, p. 80.
6. Walker, J. (1922–23). Obituary notices: Alexander Crum Brown, M.D., D.Sc., LL.D., F.R.S. *Proceedings of the Royal Society, Edinburgh, 43*, 268–276.
7. Roberts, S. (1993). *Sophia Jex-Blake: A woman pioneer in Nineteenth Century medical reform* (p. 84). (The Wellcome Institute Series in the History of Medicine). Routledge.
8. Ref. 1, Jex-Blake, p. 87.
9. Ross, M. (1996). The Royal Medical Society and Medical Women. *Journal of the Royal College of Physicians of Edinburgh, 26*(4), 629–644.
10. Thorne, I. (1951). The time in Edinburgh: 1869–1874. *Royal Free Hospital Magazine, 13*, 102.
11. Anon. (22 Nov.1870). Letter to the Editor. *The Scotsman*.
12. Cited in: M. Todd, (1918). *The life of Sophia Jex-Blake* (pp. 293–294). Macmillan and Co.
13. Bradley, J., Crowther, A., & Dupree, M. (1966). Mobility and selection in Scottish University Medical Education, 1858–1886. *Medical History, 40*, 1–24.
14. Case 138. (28 June 1873). Sophia Louisa Jex-Blake and Others, Pursuers; The *Senatus Academicus* of the University of Edinburgh, Defenders. *Cases decided in the Court of Session, Teind Court, & c. Third Series, vol. 11* (pp. 784–802). T. & T. Clark, Law Booksellers.
15. Letter. (17 July 1873). Jex-Blake, Sophia to the *Senatus Academicus*. University of St. Andrews Archives, Ref: UYUY459/Box D/Bundle 1871–73.
16. Pamphlet. (Feb. 1875) *Admission of Women to Scottish Universities*. PETITION in Favour of Mr. Cowper Temple's Bill. From Matriculated and Registered Women Students of the University of Edinburgh, to the Honourable the Commons of Great Britain and Ireland, in Parliament Assembled. 1–3.
17. Murray, J. H., & Stark, M. (Eds.). (1985). *The Englishwoman's review of social and industrial questions: 1875*. Routledge.

# Edith Pechey

# 5

*One of the 'Edinburgh Seven' mentioned in the previous chapter was Edith Pechey (married name Pechey-Phipson). Pechey was a gifted chemistry student. As such, she was the centre of an incident which was to result in significant unwelcome publicity for the University. Later in life, Pechey was one of the pioneer women doctors who practised in India while retaining strong ties to the London School of Medicine for Women.*

## Early Life

(Mary) Edith Pechey (Fig. 5.1) was born on 7 October 1845 in Langham, Essex, to Sarah Rotton, a lawyer's daughter who, unusual for a woman of her generation, had studied Greek, and William Pechey, a Baptist minister with an M.A. in theology from the University of Edinburgh [1]. Home-educated, she first worked as a governess and teacher [2].

Interested in a medical career, Pechey had hoped to take the Apothecaries' Examinations (see Chap. 1). To obtain practical experience, Pechey became indentured to Elizabeth Garrett. However, worried that other women might follow Garrett (see Chap. 1), in 1867, the Court of Examiners of the Apothecaries announced that they would no longer accept privately tutored applicants [3]. Pechey's Apothecaries route into medical studies had become firmly blocked.

After reading Sophia Jex-Blake's advertisement in a newspaper, Pechey wrote to Jex-Blake [4]:

> Before deciding finally to enter the medical profession, I should like to feel sure of success—not on my own account, but I feel that failure now would do harm to the cause, and that it is well that at least the first few women who offer themselves as candidates should stand above the average of men in their examinations.



M. Rayner-Canham and G. Rayner-Canham, *Pioneers of the London School of Medicine for Women (1874–1947)*, Perspectives on the History of Chemistry,
https://doi.org/10.1007/978-3-030-95439-0_5

**Fig. 5.1** Edith Pechey. Public domain: https://en.wikipedia.org/wiki/Edith_Pechey#/media/File:Edith_Pechey.jpg, Photo by Thomas Fall (1833–1900)

Jex-Blake added Pechey's name to those she put forward to the University Court.

Pechey moved to Edinburgh, sharing accommodation with Jex-Blake. In her book on pioneering women in medicine, Edythe Lutzker commented [5]:

> The house in which Sophia [Jex-Blake] and Edith Pechey lived soon became the meeting place for special little circles. It is hard to decide whether Sophia or Edith was more responsible for drawing them there. Sophia led the group in most matters, but she relied and even depended on Edith more than on any of the others.

## The Hope Scholarship

In addition to the joint experiences of the 'Edinburgh Seven', discussed in the previous chapter, Pechey faced a unique situation in the context of the Hope Scholarship. The Hope Scholarships had been instituted by Thomas Charles Hope. Hope had been appointed at the University of Edinburgh as the sole Lecturer in Chemistry in 1797 [6]. It was in the Spring of 1826, that Hope had offered [7]: "a Short Course of Lectures for Ladies and Gentlemen". The presence of women on campus was opposed by many academics, and the gates to the chemistry building were closed to the women. Undeterred, Hope converted a ground-floor window on a side-street into a door to enable the women to enter the building and attend his lectures.

In a letter, Lord Cockburn wrote to T. F. Kennedy [8]: "The fashionable place here now is the College; where Dr Thomas Charles Hope lectures to ladies on Chemistry. He receives 300 of them by a back window, …" The income from these chemistry lectures to women enabled him in 1828 to donate £800 (equivalent to about £80,000 now) for the founding of a University chemistry prize: The Hope Scholarship. The recipients received a monetary award, plus free use of the chemistry laboratory facilities during the following term.

## Professor Crum Brown

In Pechey's 1st year, when the chemistry marks were announced by Professor Alexander Crum Brown, she had placed third overall. The two male students above Pechey on the list were repeating the course and were therefore ineligible for the scholarship. Though the money was welcome, access to the University chemistry laboratory was even more important, as women students were barred from access to laboratories. As a result, Pechey and the other women students had to construct practical facilities in their lodgings to enable them to perform the necessary experiments.

However, Crum Brown, probably surprised by her outstanding marks, then proclaimed that Pechey was ineligible as the women students had been taught in a separate class, contradicting his earlier statements (see Chap. 3). It is appropriate to quote here Jex-Blake's own observations [9]:

> It had occurred to us that if any lady won this scholarship she might be debarred from making full use of it regards the laboratory, in consequence of the prohibition against mixed classes; but it had been distinctly ordained that we were to be subject to "all the regulations in force in the University as to examinations," it had *not* occurred to us as possible that the very name of Hope Scholar could be wrested from the successful candidate and given over her head to the fifth student on the list, who had the good fortune to be a man. But this was actually done.

Crum Brown then contradicted himself yet again by awarding Pechey a Bronze Medal of the University. This was given to the five students with the highest chemistry marks in the class. By this act, Crum Brown acknowledged that Pechey as a class member, was eligible for the medal, despite having said that, in the context of the Hope Scholarship, Pechey was not a member of The Chemistry Class.

## The Pechey Outcry

It was never mentioned anywhere as a reason for disbarring Pechey from the Hope Scholarship, but perhaps Jex-Blake was correct in concluding that the possibility of a women in the chemistry laboratory was unacceptable. Ineligibility for the Hope Scholarship was a means of avoiding this unexpected and unwelcome prospect. This explanation was suggested in a lengthy article on Pechey's case in the *Daily Review* (Edinburgh) [10].

> The only excuse that we can with the utmost stretch of charity imagine in this case would be that Dr Crum Brown thought some difficulty might arise respecting Miss Pechey's use of the scholarship (which gives free admission to the laboratory) ... but we are quite at a loss to see how any legitimate argument can be drawn thence to justify Dr. Brown in laying violent hands on a scholarship which has been fairly earned by one person for the purpose of presenting it to another.

The issue of Pechey's disqualification rapidly escalated, gaining national attention, with articles in support of her case appearing in *The Manchester Examiner and Times*, *The Spectator* ("a very odd and gross injustice"), *The Times*, *The Scotsman*, *The British Medical Journal*, and *The Lancet*. It even gained international, attention, becoming the subject of a front-page article, "Women's Rights in Scotland" in the American newspaper, *New Era* [11].

Pechey's case was appealed to the *Senatus* of the University of Edinburgh in 1870. Whereas the University Senate approved the issuing of University Certificates of Attendance to the women by a one-vote margin, by a contrary margin of one vote, the Senate denied the Hope Scholarship to Pechey.

That the Senate supported Crum Brown against Pechey, resulted in a poem titled: "A Cheer for Miss Pechey" being published in the London review magazine, *The Period*. Verses 1 and 8 are included here [12]:

> *Shame upon thee, great Edina! shame upon thee, thou hast done*
> *Deed unjust, that makes our blushes flame as flames the setting sun.*
> *You have wrong'd an earnest maiden, though you gave her honours crown,*
>
> *And eternal shame must linger round your name, Professor Brown.*
> *And I blush to-day on hearing how they've treated you, Miss P.,*
>
> *How that wretched old Senatus has back'd up Professor B.*
> *Ah! the "Modern Athens" surely must have grown a scurvy place,*
> *And the 'Varsity degraded to incur such dire disgrace.*

## Pechey Obtains Formal Qualifications

As described in the previous chapter, by 1873, the 'Edinburgh Seven' had no more avenues of appeal and they gave up hope of a qualification from the University of Edinburgh. In December 1873, Pechey wrote to the King and Queen's College of Physicians in Ireland (KQCPI) asking whether the College would admit women to the exams and grant them a Licence in Midwifery [2]. This instituted a multi-year debate within the walls of the KQCPI (see Chap. 9).

In 1875, Pechey was able to gain employment as House Surgeon at the Birmingham and Midlands Hospital for Women. The previous holder of the post, Dr. Louisa Atkins, had received her diploma in medicine, surgery, and midwifery from the University of Zurich in 1872 [13]. Atkins resigned the position to take up an appointment with the New Hospital for Women in London, run by Elizabeth Garrett. The *Birmingham Daily Post* welcomed Pechey's appointment, noting that, at the University of Edinburgh, Pechey had [14]:

> ... not only passed with credit but obtained high honours. ... In addition to her certificates, she sends testimonials from gentlemen of the highest authority in the profession, which show that she has added experience to theoretical study, ... The institution is to be congratulated on securing her services.

Over the year, 1875–1876, Pechey gained experience in abdominal surgery with Robert Lawson Taite, a recognized expert in the field [15]. Wishing to undertake additional medical studies, she travelled to the University of Bern with Sophia Clark and Ann Elizabeth Clark (see Chap. 9), who were also working with Tait. Pechey passed her medical exams, in German, on 31 January 1877, being granted an M.D. degree.

Finally, in that same month, KQCPI became the first British examining body to issue to women, Licences to Practise Medicine [16]. On 9 May 1877, Pechey sat the medical exams in Dublin and became a Licentiate of the KQCPI, a route several others would follow (see Chap. 9). In October of that year, it was as Dr. Pechey that she delivered the inaugural address at the London School of Medicine for Women (LSMW) [2].

For the next 6 years, Pechey practised medicine in Leeds, specializing in abdominal surgery. At the end of that period, she planned to move to London. However, the house she intended to occupy was not vacant for some time. Thus, in the interim, Pechey travelled to Vienna for surgical practice.

In 1883, George Alvah Kittredge [17] arrived in London from Bombay, where he attended the Annual Prize-Giving Event of the LSMW. He later reminisced about the event and aftermath [18]:

> ...I had the opportunity given me there of explaining the movement for the introduction of medical women into Bombay. I was in hopes of that the publicity thus given to our needs would bring to view the workers we required. I was almost in despair when Mrs. Dr. Garrett-Anderson, who had taken a deep interest in our scheme, and who had given me valuable advice, suggested the name of Edith Pechey, as a possible candidate for the post at Bombay. ... Mrs. Anderson thought it very doubtful if Dr. Pechey would give up her plan of settling in London, but kindly offered to write [to Pechey in Vienna] and ask her if she would consider the subject of going to Bombay. The result was that Dr. Pechey kindly met me in Paris, coming from Vienna for this purpose, ... The result was, to my great satisfaction, her acceptance of the post of Senior Medical Officer to our Fund, on a salary of Rs. 500 a month and quarters, with first-class passage out and home.

Pechey arrived in Bombay in December 1883, taking up her appointment at the new Pestonjee Hormusjee Cama Hospital for Women and Children in Bombay (now Mumbai). She lived with her companion, Ethel Dewar, as Lady Nora Scott commented [19]: "Miss Pechey is the gentleman, and Ethel the lady of the establishment."

According to Anne Crowther and Marguerite Dupree, Pechey was well able to handle life in rural India [20]:

> After a visit to a particularly inaccessible Ranee in a mountainous district, Pechey wrote home that she would not have survived without her box of fifty cigarettes. ... She reassured her correspondent that even on this solitary journey by rail and litter she had no fear of

assault. 'If anyone touches me I think they will be surprised by the force of my fist when they get it straight from the shoulder.'

Then in 1889, Pechey married Herbert Musgrave Phipson, taking the name of Pechey-Phipson. She wrote to her maternal aunt [21]:

...We have known each other so well, and worked together in so many things these five years, that there is no reason to wait for anything, and we are getting older every day. I am four years older than he is, at which I know you will shake your head, but the real objection to the marriage is that he is so unselfish that there is a great danger of my becoming a mass of selfishness... What seems more certain is that we shall be very happy together.

Pechey endeavoured to avoid any perceived cultural issue. However, in 1890, she felt she had to take issue with the practice of child-marriage in an address which was subsequently published [22]. The author of an article in the magazine, *The Queen*, commented upon her speech [23]:

The practice of compulsorily marrying girls before they had attained women's estate had been discussed from many points of view—religious, educational, political, moral, and sentimental—and the objections which had been urged were already beginning to affect public opinion. But the physiological aspect of the case had yet to be stated. It could only be stated with effect by a woman doctor. The task was a difficult one, but Mrs. Pechey Phipson carried it through with consummate ability, tact, and good feeling.

Being recognized for her efforts, Pechey was chosen as one of the two British representatives on the five-member board of the Women's Committee supporting the Age of Consent Bill in Bombay [24]. In 1891, Pechey was elected to the Senate of the University of Bombay, the first woman member of this Institution. Then in 1892, she became the vice-President of the Bombay branch of the Royal Asiatic Society, a distinguished honour, since no woman had ever been elected on the board of this élite organization. In the same year she was elected to the executive committee of the Bombay branch of the National Indian Association.

Pechey resigned from her Cama Hospital position in 1894 when diabetes began to seriously affect her health. She refused to abandon her career, however, and continued to maintain a private medical practice. When bubonic plague erupted in Bombay in 1896, she was a leader in the struggle to bring the pestilence under control. This outbreak was followed by a cholera epidemic and a widespread famine, and Pechey remained active in the forefront of the medical response to the human suffering.

Despite all of her successes, Pechey found the heat and humidity of Bombay took a toll on her health. Her Obituarist, Margaret Todd, commented [25]:

From the first, Dr. Pechey found the climate of Bombay very trying, aggravated as were its effects by hard work and heavy responsibility. After some ten years of arduous service, she retired from her post with broken health, but on the outbreak of the plague in 1896, she was among the first to organise and take part in the house-to-house visitation.

## Return to England

Early in 1905, Pechey, together with her husband, sailed from Bombay for England, spending some time along the way in Australia, New Zealand, and Canada. Her thoughts had turned to women's participation in the world of public affairs. On her return to England, Pechey was asked to represent the Women's Suffrage Association of Leeds at the third conference of the International Women's Suffrage Alliance to be held in Copenhagen in August 1906, which she accepted. Her last public appearance was as one of the leaders of the famous Mud March organized by the National Union of Women's Suffrage Societies in February 1907. It acquired the name 'Mud March' from the day's weather, when incessant heavy rain left the marchers drenched and mud-spattered [26].

Very soon after this march, Pechey underwent surgery for breast cancer, the surgeon being May Thorne, daughter of Pechey's former classmate of the 'Edinburgh Seven', Isabel Thorne. Though the operation was successful, Pechey never fully recovered and died on 14 April 1908 at her home in Folkestone, Kent. She was buried in Folkestone.

Following Pechey's death, Phipson set up a Scholarship in her name at the LSMW. Granted regularly up to 1948, when the LSMW was absorbed into the Royal Free Hospital (see Chap. 16), its terms were [27]:

> This scholarship, of the value of £40, is awarded annually in June by the Council of the London (Royal Free Hospital) School of Medicine for Women who are the trustees of the scholarship. It is open to all medical women, preferably coming from India, or going to work in India, for assistance in post-graduate study.

## Commentary

In Chap. 3, we summarized the lives of Jex-Blake and of Garrett, while in this chapter, the focus was upon Pechey. Though these were three central figures in the founding of the LSMW, they were not the only contributors to the story. In Chap. 6, we will cover the lives of the other five of the original Edinburgh Seven.

## References

1. Anagol, P. (25 May 2006). Phipson, (Mary) Edith Pechey—(1845–1908). *Oxford Dictionary of National Biography*. https://doi.org/10.1093/ref:odnb/56460. Accessed 21 Jan. 2020.
2. Lutzker, E. (1973). *Edith Pechey-Phipson, M.D.: The story of England's foremost pioneering woman doctor*. Exposition Press.
3. Hunting, P. (1998). *A history of the Society of Apothecaries*. The Society of Apothecaries.
4. Todd, M. (1918). *The life of Sophia Jex-Blake* (p. 254). Macmillan & Co.
5. Lutzker, E. (1969). *Women gain a place in medicine* (p. 63). McGraw-Hill.
6. Hirst, E. L., & Ritchie, M. (1953). Schools of chemistry in Great Britain and Ireland-VII: The University of Edinburgh. *Journal of the Royal Institute of Chemistry, 77*, 505–511.

7. Morrell, J. B. (1969). Practical chemistry in the University of Edinburgh, 1799–1843. *Ambix, 16*, 66–80.
8. Cockburn, H. (1874). *Letters chiefly connected with the affairs of Scotland* (p. 137). Ridgway.
9. Jex-Blake, S. (1886). *Medical women: A thesis and a history* (p. 82). Oliphant, Anderson, and Ferrier.
10. Anon. (1 Apr. 1870). *Daily Review*, (Edinburgh).
11. Anon. (16 Jun. 1870). Woman's rights in Scotland. *New Era*, 1.
12. Anon. (14 May 1870). A cheer for Miss Pechey. *The Period*, 12–13.
13. H. S. (1 Nov. 1924). Obituary: Louisa Atkins, M.D. *British Medical Journal, 2*, 836–837.
14. Anon. (20 Jul 1875). Editorial: Birmingham and Midland Hospital for Women. *Birmingham Daily Post*, 4.
15. Shepherd, J. A. (1986). The contribution of Robert Lawson Tait to the development of abdominal surgery. *Surgery Annual, 18*, 339–349.
16. Kelly, L. (2013). The turning point in the whole struggle: The admission of women to the King and Queen's college of physicians in Ireland. *Women's History Review, 22*(1), 97–125.
17. Anon. (1918). In memoriam: George Alvah Kittredge, M.A. *The Phi Beta Kappa Key, 3*(6/7), 308.
18. Kettredge, G. A. (1889). *A short history of the "Medical Women for India" fund of Bombay* (pp. 14–15). Education Society's Press.
19. Scott, N. (1988). *An Indian Journal* (p. 58). Radcliffe Press.
20. Crowther, M. A., & Dupree, M. W. (2007). *Medical lives in the age of surgical revolution* (p. 163). Cambridge University Press.
21. Letters of Mrs. Mary Edith Pechey-Phipson M.D., 1884–1892. Edinburgh University Library (Special Collections).
22. Pechey Phipson, Mrs. (1890). *Address to the Hindoos of Bombay on the subject of child-marriage*, Bombay Gazette Steam Printing Works.
23. Anon. (7 Mar. 1891). Mrs. Pechey-Phipson, M.D. *The Queen*, 122.
24. Anagol-McGinn, P. (1992). The age of consent Act (1891) reconsidered: Women's perspectives and participation in the child-marriage controversy in India, *South Asia Research, 3*(2), 100–118.
25. Todd, M. (1908–1909). Dr. Edith Pechey Phipson. *Magazine of the Royal Free Hospital and London School of Medicine for Women*, 883–884.
26. Anon. (30 May 2021). Mud March (suffragists). https://en.wikipedia.org/wiki/Mud_March_(suffragists). Accessed 7 Mar. 2021.
27. Anon. (1916). Dr. Edith Pechey Phipson post-graduate scholarship. *The Lancet, 187*, 1062–1063.

# 6 Others of the 'Edinburgh Seven'

*As we continue the story of the foundation of the London School of Medicine for Women, we first need to complete the account of the 'Edinburgh Seven.' As we describe here, there were more than seven; and some of the others who were involved in the Edinburgh saga were among the first students of the School (their biographies will be found in Chap. 9). We will provide biographical accounts of the other members of the 'original seven' and show how their lives continued to interact with those of one or more of Garrett, Jex-Blake, and Pechey.*

## Changes in the Edinburgh Seven

The name the 'Edinburgh Seven' has become iconic in the saga of the admission of women to the University of Edinburgh [1]. However, there were more than seven. We cannot rely on Sophia Jex-Blake for clarification. Anne Crowther and Margaret Dupree explained the problem [2]:

> One problem is seeing the medical women clearly is the strong attitude taken by Jex-Blake herself. Passionate in her feminine friendships, unwavering in her ambitions, she tended to write out of her account those who had failed to meet her standards. She quarrelled vigorously with the ELEA [Edinburgh Ladies Educational Association], in spite of their assistance to the medical women. She also disputed with Elizabeth Garrett, one of the two women on the Medical Register, over whether the women's campaign should be for the right to qualify in Edinburgh, or simply for the right to attend classes leading to a qualification abroad.

Crowther and Dupree have also addressed the fluidity of the numbers [3]:

> Later narratives, following Jex-Blake, have tended to present the women as a tiny, beleaguered minority. However, the group was somewhat larger than this. Matriculation records show ten women, not five, signing the Edinburgh register for the winter session of 1869,



M. Rayner-Canham and G. Rayner-Canham, *Pioneers of the London School of Medicine for Women (1874–1947)*, Perspectives on the History of Chemistry,
https://doi.org/10.1007/978-3-030-95439-0_6

and three more in the summer of 1870. Emily Bovell was the only woman to matriculate in the winter term of 1870, but three others came the following summer.

Three of those who attended classes at Edinburgh but with the intent of seeking a medical degree in Europe, as Garrett had done, were Anna Dahms, Annie Reay Barker, and Rose Anna Shedlock. Jex-Blake never mentioned them in her historical record [2], even though they stood alongside Jex-Blake at the later legal challenges in Edinburgh (see Chap. 4). Sophie Almond has contended that, despite Dahms, Barker, and Shedlock, being part of the 'group', Jex-Blake 'erased' them from the record as the three had 'betrayed' her. Instead of joining Jex-Blake in London and registering for the first class at the London School of Medicine for Women (LSMW), the trio travelled to the University of Paris to complete an M.D. there. Though Barker [4] and Dahms did complete an M.D. in Paris, sadly, Shedlock died of tuberculosis in Madeira in October1879, before completion of her degree [5].

Failure to complete a degree was another reason for erasure or marginalization. As an example, Sophy Massingberd-Mundy had been a favourite of Jex-Blake, but she never appeared in Jex-Blake's accounts of life in Edinburgh. In Jex-Blake's description of the Surgeons' Hall riot, she noted [6]: "S.M.M.'s great amusement" at another member of the group being scandalized by the very bad language of the mob and added [6]: "S.M.M. said she got hit." Crowther and Dupree described [7]: "Sophy Massingberd Mundy, who was pursued in the streets by male students shouting 'Whore' at a fraught period in the history of the medical women, also largely disappeared from the [Jex-Blake's] record, as she did not go on to qualify." In her biography of Jex-Blake, Margaret Todd footnoted [8]: "Miss Massingberd Mundy was one of the junior students who did not go on to graduation, but her gaiety and humour made her a real acquisition to the little circle in trying days."

Others who studied with Jex-Blake, and subsequently completed medical degrees in Europe, were: Mary Cadell, Jane Hume Claperton, Ann Elizabeth Clark, Alexina Ker, Alice Ker, Elizabeth Ker, Jane Russell Rorison, Edith Shove, Elizabeth Vinson, and Elizabeth Walker [9]. Over the following years, the numbers continued to grow. Crowther and Dupree noted [3]: "By 1873, thirty-nine women had joined the Edinburgh queue" with most likely more to follow. It was the opinion of Crowther and Dupree that it was the rapidly growing numbers which was the major cause of the opposition to the women medical students.

However, in this chapter, we will focus upon the lives of the 'Edinburgh Seven' as defined by historian, Mary Ann Elston (the lives of some of the others will be found in Chap. 9) [1]:

> … specifically the group in whose names a petition for admission to clinical instruction at Edinburgh Royal Infirmary was submitted in 1870. They were not the only women to matriculate to study medicine at Edinburgh in 1869 and 1870. But they were the first and were all committed to qualifying in medicine from the outset, whereas some of the other female matriculants had probably enrolled as partial students in a gesture of solidarity.

## Marriage

First, though, we will look at an issue which dogged the Edinburgh group: Marriage. There was no greater sin in the eyes of Jex-Blake than that of abandonment of the 'cause' of women's medical education for marriage. She forgave her intimate friend, Isabel Thorne as she had married before taking up the 'cause' [7]. Two of the original seven married but continued their studies: Matilda Chaplin (Mrs. Ayrton) and Mary Anderson (Mrs. Marshall). Mary Anderson was the sister of Elizabeth Garrett's husband.

A third member of the original group, the widowed Helen Evans (Mrs. De Lacy Evans), remarried, this time to Alexander Russel. Though Evans, herself, was 'lost to the cause', her new husband was Editor of *The Scotsman* newspaper. He proved to be a powerful ally to the cause of women's medical education through his emphatic Editorials of support.

In her book on the history of women in medicine, Edythe Lutzker commented [10]: "Sophia [Jex-Blake] received a letter at this time from a woman who urged, 'I do hope you and Miss Pechey will remain firm to the end, for really three marriages within six months is quite alarming!'" When one of the 1871 entrants, Anne Jane Anderson, became engaged to be married, it was Edith Pechey who was tasked to break the news to Jex-Blake. Pechey, who herself was to later marry (see Chap. 5), responded to Anderson [11]:

> That one should burden oneself with a husband whilst undertaking the duties of a profession is a step which seems to me foolish in the extreme, but to give up the profession for the sake of a husband is a thing so totally incomprehensible that I put it aside as one of those things which is beyond human intelligence to fathom.

The marriage of Elizabeth Garrett in 1871 (see Chap. 3) had been considered of sufficient significance that it was discussed in an editorial in the *British Medical Journal* [12]:

> It is announced that Miss Elizabeth Garrett MD is about to marry. The problem of the compatibility of marriage with female medical practice which has been much discussed will thus be partially tested. In a great city, however, and under otherwise favourable circumstances, it will no doubt offer less difficulties than in smaller towns and rural districts, or where the ordinary hardships of general practice are to be encountered. Comparatively few ladies can hope to attain the exceptional mental and material advantage which Miss Garrett enjoys.

In fact, Garrett continued with her medical practice, even after giving birth to three children: Louisa; Margaret; and Alan. Nevertheless, it had caused her to have concerns that the Leader was herself abandoning the 'cause'. In her daughter's memoirs, Garrett had confessed [13]: "… a dread lest I may be choosing my own happiness at the price of the duty I owe to women who need something which I as one of the leaders can give them."

## Isabel Pryer (Mrs. Thorne)

The first to apply to join Jex-Blake in the application for admission to the University of Edinburgh had been Isabel Thorne (Fig. 6.1). As Isabel Jane Pryer, she was born in London on 22 September 1834, daughter of Thomas and Isabel Pryer, her father being a solicitor [14]. She was educated at Queen's College, London.

In 1856, Isabel Pryer married Joseph Thorne, a tea merchant in China, and she spent her early married life in Shanghai. Partly because of the death of one of her children, Isabel Thorne had become convinced of the need for women doctors, especially to practice in China and India. Her husband fully supported her desire for a medical education.

On their return to England, Thorne attended classes at the short-lived Ladies' Medical College in London (see Chap. 8), an institution opened in 1864 to provide midwifery training for women [15]. Thorne remarked of the experience [16]:

> My experience with the Ladies' Medical College caused me to realize very forcibly the risk of imperfectly trained persons being expected by the public to undertake the duties of fully qualified practitioners, and the danger of their being called in to treat all the ills that flesh is heir to, because they were acquainted with the normal needs of childbirth, and could attend midwifery cases.

As described in Chap. 1, Thorne had then passed the Apothecaries' Examination, but had been barred from a medical qualification. However, she was not ready to give up. Thorne wrote [16]:

> In the Spring of 1869 I was on the point of starting for Paris where there was a possibility of medical classes being open to women, when I met Miss Jex Blake … I decided to join Miss Jex Blake, as also my friend Miss Chaplin, whom I had known at the Fitzroy Square

**Fig. 6.1** Isabel Thorne. https://www.hpcbristol.net/visual/vh01-004, Historical Photographs of China, Shanghai 1850s. Bath Royal Literary & Scientific Institution: L03826-003a, permission for reproduction granted

> College [Ladies' Medical College], Miss Pechey, and Mrs. De Lacy Evans, so that we were a band of five women instead of one.

Thorne's husband had returned to Shanghai by this date. Her daughter, May Thorne, recalled the travel north to Edinburgh [17]:

> My mother gave up our comfortable home … and gallantly started off with her family of four children and their nurse to endeavour with a few others to open a new career for women. My sister was 10, I was 8, my elder brother $2\frac{1}{2}$ and the younger brother 14 months old. … At the time of our journey to Edinburgh we four children had whooping cough and I understand we had a railway carriage to ourselves without any trouble, as no one wanted to travel with us! There were no regulations in those days restricting our movements on account of infectious illnesses, and off we went from King's Cross to Edinburgh.

She also recalled their first residence in Edinburgh [17]:

> … the chief thing I remember was that numerous mice ran up and down the curtains and over our beds every evening until we went to sleep, and doubtless afterwards, too. However, the mice did us no harm, and after a few evenings we thought no more of them.

However, according to May Thorne, her mother decided to find a location elsewhere [17]: "… a roomy two-storied flat at the top of a fine old house, … Miss Jex-Blake and Miss Pechey, who also wanted to study medicine, had taken a flat on the ground floor."

Having been unsuccessful in gaining a medical degree at Edinburgh, Isabel Thorne moved back south to London. There, as we discuss in Chap. 8, she was asked by the other founders of the London School of Medicine for Women (LSMW) to become Honorary Secretary of the LSMW and hold a seat on the LSMW Executive (despite her lack of a medical qualification). Thorne was the organizer and mediator in the early years of the LSMW. Without her, it is likely that the LSMW would have collapsed after a few years. Thorne fulfilled these roles from 1877 until 1908, to be succeeded by her daughter, the surgeon, May Thorne [14].

Isabel Thorne died in London in 1910. The Obituary for her in the *British Medical Journal* was fulsome in its praise [18]:

> There has passed away this week one [Isabel Thorne], who though not a member of the medical profession, did much to smooth the path of many medical women and to promote the success of a movement which has since proved so successful that its early difficulties are almost forgotten. … Together with some of her former fellow-students, she took steps which eventually led to the starting of a school of medicine for women in London, and this school, now so well known under the title of the London (Royal Free Hospital) School of Medicine for Women, Mrs. Thorne continued to serve for so long a period as thirty years. For such a post Mrs. Thorne was eminently and happily fitted, for besides being steadfast in purpose and possessed of sound judgement, she had the faculty of disarming hostility. To her long co-operation much of the great success of the school is due, …

## Matilda Chaplin (Mrs. Chaplin-Ayrton)

The second to join the 'Edinburgh Seven' was Edith Pechey (see Chap. 5) followed by Matilda Chaplin. Matilda Charlotte Chaplin was born on 22 June 1846 at Honfleur, France, daughter of solicitor John Chaplin and Matilda Ayrton [19]. She initially trained as an artist, but then she, like Isabel Thorne, enrolled at the Ladies' Medical College in London. Chaplin passed the preliminary examination for the Licence of the Society of Apothecaries just before the Society closed its professional examinations to candidates who had not attended regular medical Schools. Then Chaplin moved to Edinburgh to register as a student at the University.

In 1871, Chaplin married her cousin, the physicist and engineer William Edward Ayrton. Two years later, Ayrton was appointed Professor of Physics and Telegraphy at the Imperial Engineering College in Tokyo, and Chaplin accompanied him to Japan. Before leaving England, Chaplin had obtained a certificate in midwifery from the London Obstetrical Society and, on arrival in Tokyo, she started a School for Japanese midwives, lecturing there herself with the aid of an interpreter.

Chaplin became ill in 1877, and she and her young daughter Edith Chaplin Ayrton returned to London [20]. She resumed her medical education at the now well-established LSMW. However, Chaplin was one of the group who decided to complete her medical training at the University of Paris, which she did in 1879. The following year, she was awarded a Licentiate of the King and Queen's College of Physicians in Ireland (KQCPI). Chaplin then established a medical practice in Sloane Street, London, while commencing a study of eye diseases at the Royal Free Hospital.

Unfortunately, by 1880, Chaplin's health deteriorated further as a result of tuberculosis. In an attempt to recover, she spent a winter in Algiers and then in Montpellier, France, continuing her medical studies in both places. Sadly, Chaplin died of the disease in 1883, age 37.

## Helen Carter (Mrs. De Lacy Evans, Later Mrs. Russel)

The fourth to join was Helen Evans [21]. Born as Helen Carter in Athy, Kildare, Ireland in 1834, she was one of seven children of Helen Grey and Major Henry Carter of the 73rd Regiment Bengal Native Infantry. The Carter family moved to India, to Simla, the summer capital of British India. It was there, in 1854, that Helen Carter married cavalry officer Henry John Delacy Evans, of the Bengal Horse Artillery Regiment. Their young daughter, Helen, died in 1857, followed shortly after by the death of her husband.

By 1869, as Helen De Lacy Evans, she travelled from India to Edinburgh to join the group of seven. Then in November 1871, she married Alexander Russel who was the Editor of *The Scotsman* newspaper (though she continued to be known as Mrs. Evans in later life). Under Russel's direction, *The Scotsman* had championed the cause of Jex-Blake and the other women medical students. After marriage,

Carter ceased her studies. Despite abandoning the 'cause', she did not lose the friendship of Jex-Blake. In July 1876, Russel died suddenly from a heart attack, leaving Carter with three children and no means by which she could resume her studies.

During 1900 and 1901, Evans was a vice-President of the Committee of the Edinburgh Hospital and Dispensary for Women and Children. The Hospital, founded by Jex-Blake (see Chap. 3), was the only one in Edinburgh to provide medical and surgical care to women by women doctors. Evans died in St. Andrews, Scotland, in 1903, following a surgical procedure, age 69.

## Mary Adamson Anderson (Mrs. Marshall)

Mary Adamson Anderson was born at Boyndie, Banffshire, in 1837, one of ten children of the Reverend Alexander Anderson and his wife, Mary Gavin [22]. Her mother died when she was young. Mary Anderson not only had to take charge of the household but also the 'domestic side' of the large residential boy's School run by her father. Only after her father married again did she feel free of the responsibilities.

In 1869, Anderson moved to Edinburgh and joined Jex-Blake's 'Edinburgh Seven' obtaining distinction in several of her classes. Two years later, she married Claud Marshall, a solicitor. Sadly, her husband died a few months later, and her son lived only a few days after birth. It was after these tragedies that Anderson moved to London and registered for the first class at London School of Medicine for Women (LSMW) [23].

Unable to complete an M.D. at the LSMW, Anderson moved to the University of Paris. With the assistance of the British Ambassador, she was admitted to the Académie de Médecine, from which Garrett had earlier obtained her M.D. [24]. When she was being awarded her M.D. degree in 1882, the vice-President of the Académie de Médecine pronounced [25]:

> … but there is reason why I wish to offer [Anderson] my congratulations. For several years there have been many discussions whether women could be admitted to medical school … You have established the right for women to study medicine … You have contributed to the solution of a major social question.

Anderson established a practice in London, and she was also on the staff of the New Hospital for Women. In 1895, she moved to Cannes, France and established a practice there. Having been in ill-health for some years, Anderson returned to England and died of pneumonia at Watford in August 1910.

## Emily Bovell (Mrs. Sturge)

Emily Bovell was born in London, on 21 February 1841, the daughter of John Roach Bovell, of no occupation though with slave-owning interests in British

Guiana, and his wife, Sarah Louisa Jones [26]. Following her father's death, her mother became superintendent of a boarding house for students at Queen's College, Harley Street, providing Bovell with a possible link to Jex-Blake.

Although credited as one of the 'Edinburgh Seven,' Bovell's name is absent from the 1869 University of Edinburgh matriculation records, and from the Class Prize lists for the 1869/70 academic year, though the other women students are listed there. In 1873, Bovell moved to the University of Paris to continue her studies, when it was no longer possible to continue at Edinburgh. She eventually qualified as a doctor in Paris in 1877.

It was in Paris in 1877, that Bovell met the physician, William Allen Sturge. They returned to London, marrying on 27 September. Thereafter they set up a practice together in Wimpole Street. Bovell renewed her connection with Queen's College, Harley Street, lecturing on physiology and hygiene. In addition, she was listed as one of the three Medical Attendants for the New Hospital for Women, the others being Elizabeth Garrett and Louisa Atkins.

Her husband was a strong supporter of her professional career, and the cause of women's medical education in general. In recognition of Bovell's contribution to the medical profession, in 1880 she was nominated by the French Government for the *Officier d'Académie*, an award rarely conferred on women.

Interestingly, Bovell was not a supporter of women's political rights, as this comment in her Obituary in the *British Medical Journal* made clear [27]:

> For the so-called political rights of her sex she cared very little, but few could plead so eloquently for the breaking down of all barriers to the intellectual development of women. For every social improvement also she had a deep sympathy, and the last paper she read before the International Congress of Hygiene was on the education and maintenance of pauper children, supporting strongly the system of boarding such children in families.

In 1881, in consequence of her 'lung-mischief' (probably tuberculosis), Bovell and her husband gave up their practice in London and moved to Nice, France. Her health improved for a while, and she established her own practice there. As the first woman doctor in Nice, she gained a good number of female patients. Bovell also campaigned on public health issues, such as improving the sewer system. She died at Nice in April 1885.

## Commentary

You, the reader, have now been introduced to Jex-Blake and Pechey, together with the other members of the 'Edinburgh Seven'. These were not the only women of the time who had decided that medicine was to be their life-long ambition. There were many seeking a pathway into medical School. Before we continue with the next part of the account, that of the formation of the London School of Medicine for Women, we will digress to examine the reason why medicine was seen as an acceptable career for middle-class Victorian girls. In particular, we will show the influence of the print media on providing the stimulus.

## References


1. Elston, M. A. (28 May 2015). Edinburgh seven. *Oxford Dictionary of National Biography*. https://doi.org/10.1093/ref:odnb/61136. Accessed 27 Aug. 2021.
2. Crowther, M. A., & Dupree, M. W. (2007). *Medical lives in the age of surgical revolution* (pp. 41–42). Cambridge University Press.
3. Ref. 2, Crowther & Dupree, p. 39.
4. Almond, S. (2020). The forgotten life of Annie Reay Barker, M.D. *Social History of Medicine, 34*(3), 828–850.
5. McIntyre, N. (2016). The fate of Rose Anna Shedlock (c1850–1878) and the early career of Émile Roux (1853–1933). *Journal of Medical Biography, 24*(1), 60–67.
6. Todd, M. (1918). *The life of Sophia Jex-Blake* (p. 292). Macmillan and Co.
7. Ref. 2, Crowther & Dupree, p. 42.
8. Ref. 6, Todd, Footnote 98.
9. Ref. 2, Crowther & Dupree, p. 41.
10. Lutzker, E. (1969). *Women gain a place in medicine* (p. 96). McGraw-Hill.
11. Letter, Edith Pechey to Anne Jane Anderson, (3 Sept. 1873). Edinburgh University Library, Gen 2213/10.
12. Anon. (7 Jan. 1871). Notes. *The British Medical Journal, 1*, 16.
13. Garrett Anderson, L. (2016). *Elizabeth Garrett Anderson, 1836–1917* (p. 177). Cambridge Library Collection, History of Medicine.
14. Anon. (9 May 2021). Isabel Thorne. https://en.wikipedia.org/wiki/Isabel_Thorne. Accessed 29 Jul. 2021.
15. Molinari, V. (2011). Schools of their own—The Ladies' Medical College and London School of Medicine for Women. In D. S. Andréolle & V. Molinari (Eds.), *Women and science, 17th century to present: Pioneers, activists and protagonists. Part II* (pp. 99–124). Cambridge Scholars Publishing.
16. Thorne, I. (1905). *Sketch of the foundation and development of the London School of Medicine for Women* (pp. 4–5). G. Sharrow.
17. Thorne, M. (1951). Then and now. *Royal Free Hospital Magazine, 13*(32), 128–131.
18. Anon. (15 Oct. 1910). Obituary: Isabel Thorne. *The British Medical Journal, 2*, 1198.
19. Elston, M. E. (8 Oct. 2020). Ayrton [née Chaplin], Matilda Charlotte. *Oxford Dictionary of National Biography*. https://doi.org/10.1093/ref:odnb/949. Accessed 27 Aug. 2021.
20. Orme, E. (1883). Matilda Chaplin Ayrton, M.D. *Englishwoman's Review, 14*, 343–350.
21. Anon. (2 Dec. 2020). Helen Evans. https://en.wikipedia.org/wiki/Helen_Evans. Accessed 21 Aug. 2021.
22. Anon. (11 Dec. 2020). Mary Adamson Anderson Marshall. https://en.wikipedia.org/wiki/Mary_Adamson_Anderson_Marshall. Accessed 27 Aug. 2021.
23. Anon. (20 Aug. 1910). Obituary: Mary Adamson Marshall, M.D. Paris. *British Medical Journal, 2*, 498.
24. Reynolds, S. (2007). *Paris-Edinburgh: Cultural connections in the belle epoque* (p. 179). Ashgate Publishing.
25. Anderson, G. (27 Aug. 1910). The late Mrs. Mary Marshall, M.D. *British Medical Journal, 2*, 574.
26. Anon. (21 Apr. 2021). Emily Bovell. https://en.wikipedia.org/wiki/Emily_Bovell. Accessed 27 Aug. 2021.
27. Anon. (30 May 1885). Obituary: Emily Bovell Sturge, M.D. *British Medical Journal, 1*, 1131.

# 7 Women as Lady Doctors

*For the London School of Medicine for Women to have been a success, there had to be sufficient women interested in a medical career. In fact, there was considerable pent-up demand throughout Britain, as medicine had been promoted as an acceptable career option for middle-class Victorian girls. In this prelude chapter to the founding of the School, we will examine the reasons.*

## The Role of Girls' Magazines

The role of girls' magazines was incredibly important in shaping the world view of Victorian and Edwardian middle-class girls. The magazines extolled the infinite possibilities for the 'New Girl.' The three most significant of these magazines were: the *Atalanta* (1887–1898), the *Girl's Realm* (1898–1915), and the *Girl's Own Paper* (1880–1927) [1].

Though the viewpoint of the girls' magazine depended upon the editor of the time, these magazines were very progressive and quite adult in their messages. The issues contained a significant proportion of intellectual material, including articles on science topics; reviews of Colleges and universities for young women; and the widening career opportunities for women. With the girls' magazines urging them forth, many of the young girls of the 1880s to 1920s saw it almost a duty of their generations to see a career in their future, not endless domesticity.

The *Girl's Realm* was the most forward-thinking of all, contending that girls had an entitlement to an equal education (and career opportunities) to that of their brothers. In a study of the *Girl's Realm*, Kristine Moruzi explained why it was the most progressive of the girls' magazines [2]:



M. Rayner-Canham and G. Rayner-Canham, *Pioneers of the London School of Medicine for Women (1874–1947)*, Perspectives on the History of Chemistry,
https://doi.org/10.1007/978-3-030-95439-0_7

> In part because it first appeared so late in the nineteenth century – and, therefore, was unlike many other girls' periodicals of the period, which had stronger ties to the Victorian era – the *Girl's Realm* is significantly less constrained by nineteenth-century ideas of femininity. Unconcerned with a nostalgic feminine idea of the past, the magazine instead uses current events to fashion girlhood as a time of bravery and courage. The girls in its pages are educated and feminine and the Girl's Realm positions them as heroes. Involved in adventures both at home and abroad, girls are capable and confident when they need to be. ... and nonfiction articles encourage girls to seek opportunities beyond the domestic sphere.

## Medicine as a Career Option

The girls' magazines extolled medicine as an excellent career option for science-educated girls. An article by Edith Huntley, M.D., in an 1887–1888 issue of *Atalanta* opened with [3]: "The object of this paper is to commend the medical profession as a newly-opened and most promising career for women to the consideration of all young educated women into whose hands it may fall." Huntley was a graduate of the London School of Medicine for Women (LSMW) [4], and she had written a book: *The Study and Practice of Medicine for Women* [5].

An article in 1898 by Ruth Young in *Atalanta* also encouraged girls to become doctors. She pointed out that it was the 'Battle of the Eighteen Seventies' when the rights of women to study medicine were established. She concluded her commentary with a very strong statement of encouragement [6]:

> At the present time women doctors in many cases find it difficult to obtain posts as medical officers in institutions, owing, probably to a great extent, to the prejudice which the man-mind cherishes against a woman going out of her proper sphere; which proper sphere is still, alas, looked on by too many men as the frying-pan and the dust-pan; ... When the fact is recognised that it is the duty and right of every woman to make use of her gifts and opportunities, medical women will stand as good a chance as their brother professionals of gaining the posts they are perfectly capable of filling nobly.

In the pages of *The Young Woman*, a series titled "How Can I Earn My Living?" devoted the first article to becoming a doctor. Miss Billington makes the case that a "good all-round schooling" was most definitely not the way to go [7]:

> And I have intentionally opened this series with a consideration of the medical profession and its closely-allied calling of nursing, for it points a lesson upon which I cannot more strongly insist, and that is that the girl who intends to fight her own battle in the world must be prepared for the combat with a *specialised* education.

Enthusiasm for the potential of a medical career was also expressed in the pages of *Girls' Realm*. This magazine published two articles on medical careers for girls: one in 1900–1901 by Margaret M. Traill Christie, M.D., B.S., D.P.H., [8]; Christie reviewed the career opportunities for a woman doctor in 1900 [8]:

> There is scientific research and medical literacy work, and posts are open as lecturers, tutors, and demonstrators in women's medical schools. Nearly half the women who qualify do so

> with a view to the high service of medical missions abroad. Women doctors go mostly to lands where the native women refuse or dislike the attendance of men as doctors. … Several Poor Law infirmaries, Fever Hospitals, and Lunatic Asylums have one or more women on their staff. … Several hospitals have women as house surgeons, clinical assistants, or anæsthetists. These posts are usually unpaid, but are sought after for the sake of experience and honour.

Ten years later, addressing the next generation of young women, Elizabeth Sloan Chesser, M.B. provided a very optimistic view [9]: "The girl who desires to study medicine has very little opposition to contend with to-day. The day when active resistance to medicals is past; they enjoy equal opportunities with men in the study and practice."

At the end of the article, Chesser addressed the subject of career options [9]:

> And when the goal is reached, and a girl graduates in medicine, what of her prospects? Compared with other occupations for women, medicine stands high. Medical women hold hospital and dispensary appointments in various institutions in this country; they act as assistants, and are eligible for asylum posts. In connection with medical inspection of schools and preventative medicine generally, there is a large field for their energies. A great many women are engaged in lucrative practices all over the British Empire. In the East, in India and China especially, medical women are doing splendid work in connection with medical missions. Medical posts under parish councils and such appointments as those in connection with bacteriological laboratories may be held by women.

We will return to the question of women doctors travelling to the East in Chap. 9.

## What to Do with Our Girls

Even in the 1880s, parents were becoming concerned about available careers for their daughters. In 1884, a book by Arthur Vanderbilt was published with the title: *What to do with Our Girls; or, Employment for Women* (previously mentioned in Chap. 2). In the introduction, Vanderbilt explained [10]:

> No question is of probably greater interest than that of "What to do with our girls?" so as to enable them-in case of necessity- either to add to an income or earn a livelihood. It is indeed of the more importance, inasmuch as it is an acknowledged fact that, though we are admittedly the wealthiest nation in the world, yet there are in our midst hundreds and thousands of women, gently and carefully nurtured, but dependent upon their own exertions for subsistence, and often in the direst state of destitution and want, through their utter inability to find, or to obtain, really remunerative and suitable employment. It is therefore with the hope of helping these, that this little book has been written.

Vanderbilt recommended medicine as a career for young women with a science background. He first suggested that 'Medical Women' had a role in Asia [11]: "In the East, where a male doctor is seldom, if ever, allowed to enter the Harems or Zenanas, they have an immense field for usefulness, which even the most prejudiced will scarcely deny." Vanderbilt also foresaw very specific roles for 'Medical Women' in Britain [11]:

> There are many manufactures of a dangerous and unhealthy nature in which women are employed, such as the white lead works, &c., and, without doubt, a lady doctor would find no difficulty in obtaining an appointment in connection with them. … There is also a fine field for Medical Women in the mining and mountainous districts of Wales and Scotland. The colliers' wives, and peasantry generally, do not, as a rule, care for doctors, and even in most serious illnesses prefer to do without them.

## Effect of the First World War

In 1915, the Editor of *The Girl's Realm* pointed out that planning for a career was an absolute necessity for the teen girl reader. The possibility of marriage was an unlikely prospect, considering the enormous number of male deaths on the battlefields, particularly among the middle-class officers and men [12]: "I address these remarks principally to girls who are facing life at the opening of their careers. The war has made us all face reality, and, not least of all, the [young] women of the world."

The same year, the feminist activist Elizabeth Sanderson Haldane read a paper to the Educational Science Section of the British Association, published in *The School World*, with the same concerns, though in more detail. She pointed out that the War had opened up whole new avenues of employment to young women. Like the Editor of *The Girl's Realm*, Haldane emphasized that marriage was unlikely to be an option for most girls [13]:

> We know too well that there must be a shortage of men with casualties reaching very many thousands. Marriage will cease to afford the normal career for very many women, and far more women will have to obtain economic independence. Parents will have to think much more seriously of the question: "what to do with our daughters." Even now the ordinary middle-class father is anxiously applying himself to the problem, and in quite a different spirit from that of earlier days.

## Commentary

In this chapter, we have shown that becoming a 'lady doctor' was not the preserve of the few, such as the 'Edinburgh Seven.' Instead, from the 1880s onwards, medicine was widely promoted as a career for scientifically oriented young women. The medical establishment had good cause to believe that after 'the few', would come 'the many'. It was to satisfy this need for a medical pathway for women that Jex-Blake, Garrett, Thorne, and others, decided that the only option was to found a separate School of medicine for women. Its tortuous and complex founding is the focus of the next Chapter.

## References

1. Dixon, D. (2001). Children's magazines and science in the Nineteenth Century. *Victorian Periodicals Review, 34*(3), 228–238.
2. Moruzi, K. (2009). Feminine bravery: The *Girls' Realm* (1898–1915) and the Second Boer War. *Children's Literature Association Quarterly, 34*(3), 241–254.
3. Huntley, E. (1887–1888). Employment for girls. Medicine. *Atalanta, 6*, 596.
4. Digby, A. (1999). *The evolution of British general practice, 1850–1948* (p. 164). Oxford University Press.
5. Huntley, E. A. (1886). *The study and practice of medicine for women*. Farncombe & Co.
6. Young, R. (1898). The medical profession as a calling for women. *Atalanta, 11*, 694–696.
7. Billington, Miss. (Nov. 1893). How can I earn my living? I—As a doctor. *The Young Woman*, 61.
8. Christie, M. M. T. (1900–1901). Careers for girls. IX.-Medicine. *Girls' Realm, 3*(1), 163–168.
9. Chesser, E. S. (1909–1910). Careers for girls. I.-Medicine. *Girls' Realm, 12*, 329–330.
10. Vanderbilt, A. T. (1884). *What to do with our girls; or, employment for women* (p. 1). Houlston & Sons.
11. Ref. 10, Vanderbilt, p. 85.
12. The Editor. (1915). The war and women: The influence of the world-conflict on women's status and work. *The Girl's Realm, 17*, 45–46.
13. Haldane, E. S. (1915). The vocational education of girls after the war. *The School World, 17*(203), 401–404.

# The Founding and Early Years of the LSMW

# 8

*For the Edinburgh Seven, and those who subsequently joined them, having no more avenues of appeal, the struggle for a medical education in Scotland came to an end. There were two other options: to apply to a medical School in another country which admitted women; or to attempt to found a medical School in London and find a hospital which would provide the opportunity for clinical experience. It was Isabel Thorne and Sophia Jex-Blake decided on the latter course of action.*

## The Female Medical Society

Before commencing a discussion on the London School of Medicine for Women (LSMW), it is important to note that there had been an earlier body in London for the education of women in medicine: the Female Medical Society (FMS). The FMS was inaugurated in 1862 with 43 patrons from almost every branch of public life [1]. While, as its name implied, the Society hoped ultimately to secure the admission of women to the medical profession, it chose to concentrate first on midwifery.

Following from these plans, in 1864, classes were offered under the name of the Ladies' Medical College. Percy Edmunds, son of the Honorary Secretary of the FMS, felt obliged to correct a statement in the *British Medical Journal* that the LSMW had been the first medical School for women in Britain. Edmunds stated that the Ladies' Medical College had offered a rigorous programme of training [2]:

> The course on obstetrics alone comprised eighty lectures of one hour each; and each student was required to attend personally twenty-five deliveries under qualified superintendence at a lying-in hospital or maternity charity. Other courses comprised anatomy and physiology, chemistry, materia medica, diseases of women, diseases of infants, and "general medical science."



M. Rayner-Canham and G. Rayner-Canham, *Pioneers of the London School of Medicine for Women (1874–1947)*, Perspectives on the History of Chemistry,
https://doi.org/10.1007/978-3-030-95439-0_8

To ensure that access to a hospital was feasible, in 1867, the FMS affiliated with the British Lying-in Hospital. In this way, the Hospital enjoyed the advantage of the free services of the Society's obstetrical students while the students had access to clinical practice.

Though the plan had been for the College to ultimately become a complete medical School for women, little progress was made. An editorial in the *British Medical Journal* of 1871 commented [3]:

> Many well-wishers of women have, in fact, not shown it any favour; and many sensible women desiring a complete medical education, such as Miss Garrett and the *septem contra Edinam*, have practically testified their opinion by keeping away from it. The main ground of disfavour is, we suppose, that it is not a medical college in the proper sense of the term. A medical college must fulfil certain conditions which all male medical colleges observe, and on which depends their recognition by the examining bodies, and their value as educational bodies.

In fact, accepting reality, in 1872, the College re-named itself the Obstetrical College for Women. An appeal for funds in the Winter of 1872–73 met with insufficient support and the College closed shortly afterwards.

## The Beginnings of the LSMW

Though the *British Medical Journal* was critical of the FMS itself, it was very supportive of a medical School for women. In the same article as above, it made its opinion clear [3]:

> They [women] must establish their own schools. It is quite evident from the number of eminent physicians who have testified their sympathy with the past efforts of ladies, that this rests entirely with themselves. They will be readily enough admitted to obstetrical hospitals; and, if they desire the use of a clinical hospital, they must find, or found, one not pre-occupied by male students; both can be done.

In 1874, Isabel Thorne and Sophia Jex-Blake contacted sympathizers and assembled a Provisional Council. Of the 21-member Council, the only women were Elizabeth Blackwell and Elizabeth Garrett Anderson. Funds were raised for the purchase of a property on Henrietta Street in London. They were then able to find volunteer Lecturers from hospitals across central London: from St. Mary's Hospital; University College Hospital; Charing Cross Hospital; Guy's Hospital; Westminster Hospital; and Middlesex Hospital [4].

Whereas the *British Medical Journal* was supportive of the training of women doctors, *The Lancet* was strongly opposed to the formation of the School. One of its criticisms was the claimed low-calibre of the lecturers. In an 1874 column, 'Medical Annotations,' the anonymous author expressed his negativism [5]:

> It is true that none of the lecturers … have been distinguished as teachers of the first rank, and many are probably not known beyond the narrow circle of their own small school. Nevertheless, these gentlemen serve a purpose, and do after all constitute a staff of lecturers,

**Fig. 8.1** The London School of Medicine for Women ca. 1876. Public domain: London Metropolitan Archives. Photograph by Miss Ley-Greaves, ca. 1876

> and will no doubt acquit themselves to the best of their abilities should the opportunity ever present itself.

Despite this criticism, the London School of Medicine for Women (LSMW) duly opened its doors on 12 October 1874 (Fig. 8.1). Alice Rowland (Mrs. Ernest Hart), who had commenced her education by the pharmacy route (see Chap. 2) and who subsequently registered as an LSMW student, reminisced [6]:

> This house was said to have been the domicile of a favourite lady of a royal personage. It bore some traces of better times. From the ceiling of the drawing-room, which become our dissecting room, still hung the chandelier with its cut-glass pendants: and the tall French windows opened on to a shady garden, which was once gay with flowers.
> Another large room—probably the old dining-room—was used as the lecture hall, and the small rooms upstairs had various uses: but the whole place was dreary and comfortless.
> We all, however, set to work with extreme earnestness, determined to conform to every requirement exacted by the medical examining bodies, so as to qualify for a medical degree, when the closed doors of the profession should open.

## Access to Hospitals

As an independent School, the LSMW lacked any opportunity for the students to undertake clinical studies in a hospital. The 1874 'Medical Annotations' article (mentioned above) in *The Lancet* pointed out, this precluded the possibility of the women students obtaining a medical degree [5]:

> We believe that, so far, the school is not attached to any hospital in which the students could prosecute their practical and clinical studies, and complete their professional education. Yet, without a hospital, the school cannot and will not be recognised by the examining bodies. To acquire an existing hospital, or to build a new one, will be one of the greatest difficulties with which the Provisional Council will have to contend. It is, we trust, out of the question that women shall be allowed to study in hospitals connected with medical schools, which henceforth, for the sake of distinction, must be designated as medical schools for men.

To provide the required hospital facilities, Elizabeth Garrett Anderson offered her small New Hospital for Women (see Chap. 3) [7]. However, it only had twenty-six beds, not the minimum of one hundred required under the regulations for an authorized connected hospital [9]. One obvious candidate was the nearby Royal Free Hospital (RFH), which lacked a medical School of its own. Despite this shortcoming, the RFH refused to consider the possibility of having the LSMW affiliated with it.

By the beginning of 1875, there seemed little future for the LSMW as Olivia Campbell recounted [8]:

> At the start of the school's second year, five new students appeared [adding to the 23 at the end of the first year]. ... But things were looking bleak once again. They were no closer to providing the necessary local clinical training than when the school opened, and students started leaving. The school's closure felt imminent. A delegation from the London school went off to plead with the government to intervene on their behalf. ... Their plea was ignored, and it was decided that the school would have to close after the next winter session.

May Thorne, daughter of Isabel Thorne, described the despair of the students by early 1877, they seemed as far as ever from a complete medical education and a qualifying body willing to examine them [9]:

> When the three years' curriculum was completed in 1877, the thirty-four women who had taken it seemed to be as far off from ever from their goal. The voluntary contributions all the students and their friends had given had all been expended in teaching and maintenance, so no classes were held in the Summer session of 1877. The outlook was indeed very gloomy.

The saviour of the LSMW was the politician, Sir James Stansfeld, Radical Member of Parliament and social reformer. As Campbell explained [10]:

> It was around this time [January 1877] that Stansfeld had a chance meeting with the chairman of the Royal Free Hospital's board while they were both on vacation. He [The Chairman] promised to consider an association with the London school [LSMW], though he did not appear to be in awe of the staff. Stansfeld set to work talking up the school.

Stansfeld's pleading was successful, and later in 1877, the RFH Board unanimously accepted the proposal for clinical instruction of the LSMW students. However, as Campbell pointed out, attachment came with strings [10]:

> The college would have to pay them [RFH] 300 guineas a year to compensate for the potential losses brought about by the relationship to a women's institute, and give them all student fees paid for clinical instruction, promising it would amount to no less than £700 per year.

Though the RFH Board accepted the women students of the LSMW—with the conditions described above—the women students were not welcomed within the walls of the RFH. Mary Scharlieb, in her later reminiscences, recalled [11]:

> The doctors did not think that they would find in them desirable pupils, the nursing staff who had hitherto performed the duties which generally fall to medical students were much aggrieved because the entrance of the women would take from them both work and prestige, and finally the Board of Management and the other authorities dreaded the effect that their new move might have on the public mind and feared that their funds would suffer. The friends of women doctors were few, …

## The Programme of Study

With the access to the RFH, it was now possible to offer the full programme of medical education. In the early 1890s, the curriculum was as follows [12]:

**First Year**
Before Christmas: Physics and Biology, Tutorial Materia Medica.
After Christmas: Practical Pharmacy.
Throughout Winter Session: Anatomy, Practical Anatomy, and Chemistry.
Summer: Practical Chemistry. and Materia Medica.
First Examinations.
**Second Year**
Anatomy and Practical Anatomy.
Physiology.
Minor Surgery once a week before Christmas.
Out-Patient Post once a week after Christmas.
Summer: Practical Histology, Practical Physiology, Lectures on Histology.
Second Examinations.
**Third Year**
Medicine, Surgery, Hospital Posts.
Repetition of Materia Medica, Pathology.
Third Year Examinations.
**Fourth Year**
Medicine, Operative Surgery, Midwifery and Gynæcology, Mental Diseases, Forensic Medicine, Hygiene, Ophthalmic Surgery.
Hospital.
Intermediate Examinations.
**Fifth Year**
Fevers.
Hospital.
Final Examinations.

## Who Would Grant Degrees?

The final, and perhaps the greatest hurdle, was: Who would grant degrees to women graduates of the London School of Medicine for Women? Previous attempts at passing an enabling bill to give Universities the power to confer medical degrees upon women—if they wished—had failed. Nevertheless, in 1876, the Conservative M.P. Russell Gurney re-introduced such a bill [13]. This time, The Medical Act 1876 passed after lengthy discussion and with much ferocious opposition. The Act which repealed the previous Medical Act in the UK, allowed all British medical authorities to licence every qualified applicant whatever their gender. The Act obtained the Royal Assent and became law despite Queen Victoria's strong private objections to women's medical training. Though the Act enabled a University to grant medical degrees (and other degrees) to women, it did not require them to do so.

By this date, the LSMW had lost some of the pioneers. They had decided to obtain their degrees in a more hospitable environment in Europe; for example, both Sophia Jex-Blake and Edith Pechey passed their final medical examinations in January 1877 at the University of Bern, Switzerland. However, their qualifications were not recognized in Britain by the General Medical Council of Great Britain and Ireland (GMCGBI). Following from the passage of the Act, the King's and Queen's College of Physicians in Ireland (KQCPI) admitted these women to their final examinations (see Chap. 9), enabling them to be registered by the GMCGBI and to practice in Britain [9]. Nevertheless, as generous as the KQCPI had been, it could not be the long-term answer for the future generations of LSMW students.

Thus, it came as a great relief when the University of London finally agreed to admit women to their medical examinations and grant them degrees. The process was started in January 1877 by the Rev. Septimus Buss who moved the following resolution in the Convocation of the University of London [14]: "That it be referred to the Annual Committee to consider a report upon the best means of carrying into effect the desire of Convocation that the degrees of the University should be open to women." The motion was carried by 22 to 16. In June 1877, the University decided to admit women to its medical examinations and hence its medical degree.

Exhilarated by the news, and signed by 31 women (including Elizabeth Garrett Anderson, Annie Reay Barker, Fanny Jane Butler, Anna Dahms, Isabella M. Foggo, Alice M. Hart, Sophia Jex-Blake, Mary A. Marshall, Mary Edith Pechey, Rose A. Shedlock, Edith Shove, Isabel Thorne, Elizabeth Ireland Walker, Jane E. Waterston, and nearly all the others being students at the LSMW), the following Memorial was sent to the Chancellor and Senate of the University of London in June 1877 [15]:

> We the undersigned women, who are engaged in the practice and study of Medicine, have heard with the greatest satisfaction of the resolution of the Senate, to admit women to the Medical Examinations and the Degrees of the London University. The fact that a complete Medical School for Women, with the necessary hospital practice, has recently been established in London, leads us to think that the present is a fitting time for extending to women

> the incentive to wide and patient study, which is afforded by the high standard of the London degree. We believe that this incentive will prove to be in all its bearings as valuable to women as it has been to men.
> We beg therefore to tender our sincere thanks to the Senate for the action they have taken, and to express our earnest hope that the necessary steps for giving effect to their resolution will be completed as soon as possible.

## A Lookback in Time

The founding and success of the London School of Medicine for Women was recognized by the women's magazine, *Mayfair*, as a truly momentous event of the 1870s. An Editorial of 1878 reminisced upon the changes [16]:

> Eight years ago, what would have been said if a middle-aged man had predicted that he would live to see Englishwomen, with the sanction of the law, and under the friendly patronage of 'Society,' practicing for gain the most arduous, ill-remunerated, and, in some respects, most repulsive of all the learned professions? He would probably have been looked upon by charitable people as a dreamer of vain dreams or by intolerant people as a pestilent lunatic. And yet, strange as it may appear, nine short years have sufficed to evolve this extraordinary social phenomenon. ... In this current year of 1878 there are at least about a dozen scientifically-qualified doctoresses practicing the profession in different parts of England. London, Bristol, Edinburgh, Leeds, Birmingham, have all their lady physicians, and there is, we are told, one in rural practice somewhere in the wilderness of Somersetshire.

## The Final Challenge

There was one more obstacle to overcome: admission to the Royal College of Physicians of London. In 1878, only 16 Fellows of the College had supported a resolution in favour of the admission of women to the examinations for the Licence of the College [15]. By 1895, the number voting in favour had risen to 50 and the opponents, only 59. Success came in 1908, when the vote was 74 in favour and 33 opposed. The article in the *LSMW Magazine* enthused about the result [17]:

> This victory, more complete perhaps than even the most sanguine had ventured to anticipate, was gained after two prolonged and exhaustive debates, in which every aspect of the question was thoroughly discussed. ... The special thanks of medical women are due to Dr. Herringham, who has throughout shewn himself one of their most enthusiastic supporters, and who, alike by his speeches and by his personal influence and untiring work, has contributed materially to the successful issue.

## Building Expansions: 1885–1892

With the challenges behind them, the LSMW underwent steady growth which necessitated expansion of facilities. Between 1885 and 1892, these alterations were piecemeal, as described by George Mudge, who had been Head of the Biology Department of the LSMW between 1899 and 1936 [18]:

> After the lapse of eleven years, in 1885 it became necessary to put up an additional lecture room in the garden and to re-arrange some of the rooms in the house. These alterations needed the consent of the ground landlords, who raised objections to the study of Human Anatomy on their premises. These difficulties, however, were overcome. These early students of Anatomy worked under very trying conditions. The Anatomy Department was in the attic, there was nothing but the barest facilities and as there was no arrangement for heating, work during the winter must have tried the endurance of the hardiest.

To cope with even more students, a property on Hunter Street was purchased in 1891 which backed onto the School and the two buildings connected. This purchase and changes amounted to a significant cost. Then in 1892, the neighbouring property on Hunter Street was purchased. It was remarked by Mudge [18]: "There was considerable perplexity in meeting this, but as so many times before, friends of the School came to its rescue."

## Building Expansions: 1896 to 1900

In 1896, to indicate the increasingly close relationship with the RFH, the name was changed to the London Royal Free Hospital School of Medicine for Women (LRFHSMW). This name change coincided with plans by the LSMW Executive Council to substantially enlarge and modernize the facilities. Mudge recalled [18]:

> The three old houses were becoming dilapidated and the construction and arrangement of rooms that prevail in private houses, were altogether unadaptable to the purposes of a medical school. … And yet in 1896 we were faced with the disconcerting prospect of carrying on with our studies, with bricks falling all around us, with clouds of dust hovering about us, with noise and din, with inadequate heating and lighting, with scanty water supply, with the outside elements finding their way in through gaps in the defences, and certainly no prospect of morning tea and biscuits at 11 a.m. None of the students of this generation have had to endure a lecture while the bricklayers and carpenters have been carrying on with their unsilent work.

By the time the expansions were completed in 1900, the LSMW was deep in debt. Mudge commented [18]: "The clearance of this debt is a story almost equal in ingenuity and romance to that by which the finances of Old Japan passed into those of New Japan."

## Departure of Jex-Blake

The proposal for a massive building programme was initiated by the Dean of the time, Elizabeth Garrett Anderson, with full support of the Council. However, it was bitterly opposed by Sophia Jex-Blake who considered it a "reckless extravagance" and she resigned in 1896 and severed all connection with the LSMW [19].

In Margaret Todd's biography of her partner, Jex-Blake, she reviewed the interaction of the three key figures in Jex-Blake's departure. Though Todd was obviously biased, this account provides a glimpse into the personality interplay [20]:

> The truth is that S. J.-B., to the day of her death and with all her faults, was an incorrigible idealist; and Mrs. Anderson, rich though she was in excellent qualities, seemed to her to be lacking in certain capabilities of insight and imagination which outweighed everything else....
>
> It was a painful situation all round, but like so many painful situations, it called forth something fine. Mrs. Thorne was *persona grata* with all parties, and finally Mrs. Thorne stepped into the breach and allowed herself to be elected Honorary Secretary of the School. "About the best possible," wrote S. J.-B. in her diary, "with her excellent sense and perfect temper. So much better than I".

## Building Expansions: 1914 to 1916

The number of women students increased further, and by 1914, there were over 300, resulting in an urgent need, yet again, for additional facilities. Mudge explained [18]: "Under the auspices of Her Grace the Duchess of Marlborough, an appeal was issued and the new block, built on a plot of land adjoining the School on Wakefield Street, was commenced in July, 1915." An additional appeal for funds was made by the then-Prime Minister, A. J. Balfour, resulting in the total required funding being obtained.

Construction was completed in 1916, including new science laboratories. The official opening was performed by Queen Mary. The leading role played by the Duchess of Marlborough was recognized [21]:

> Part of the new half-basement is used for an extension of the physics laboratories, and bares the name of the "Consuelo Duchess of Marlborough, Physics Department," It includes a lecture theatre, lecturers' research room, and rooms specially prepared for experimental work in light.

## Commentary

In this chapter, we have surveyed the events which occurred at the founding of the LSMW. In Chap. 9, we will provide brief biographies of some of the women who joined the first intake, other than those who we discussed in Chap. 6. It was these women who were to provide the first LSMW cohort and who helped forge the identity of this unique institution.

## References

1. Molinari, V. (2012). Schools of their own: The Ladies' Medical College and the London School of Medicine for Women. In D. S. Andréolle & V. Molinari (Eds.), *Women and science, 17th century to present: Pioneers, activists and protagonists* (pp. 99–124). Cambridge Scholars Publishing.
2. Edmunds, P. J. (18 Mar. 1911). The origin of the London School of Medicine for Women. *British Medical Journal, 1*, 659–660.
3. Anon. (23 Sept. 1871). Minerva Medica, *British Medical Journal, 2*, 356.
4. Thorne, I. (1905). *Sketch of the foundation and development of the London School of Medicine for Women*. G. Sharrow.
5. Anon. (17 Oct. 1874). Medical annotations: London School of Medicine for Women. *The Lancet, 104*, 561–562.
6. Hart, E. (1924). The London School of Medicine for Women—Past and present. By a pioneer—Mrs. Ernest Hart. *Magazine of the London (Royal Free Hospital) School of Medicine for Women, 19*(89), 129.
7. Glynn, J. (2008). *The Pioneering Garretts: Breaking the barriers for women*. Hambledon Continuum.
8. Campbell, O. (2021). *Women in white coats: How the first women doctors changed the world of medicine* (p. 279). Park Row Books.
9. Thorne, M. (1951). Then and now. *Royal Free Hospital Magazine, 13*(32), 128–140.
10. Ref. 8, Campbell, p. 285.
11. Scharleib, M. (1925). *Reminiscences* (2nd ed., p. 51). Williams and Norgate Ltd. Credit: Reminiscences/[Mary Scharlieb]. Wellcome Collection. In copyright.
12. Billington, Miss. (Nov. 1893). How can I earn my living? I—As a doctor. *The Young Woman*, 62.
13. Ref. 8, Campbell, pp. 285–286.
14. Ref. 9, Thorne, p. 134.
15. London's Pioneers in their Own Words: Memorial of Women Doctors. (2019). https://archive.senatehouselibrary.ac.uk/exhibitions-and-events/exhibitions/rights-for-women/resources/letters-and-transcripts/memorial-of-women-doctors. Accessed 10 Sept. 2021.
16. Anon. (27 Aug. 1878). Editorial. *Mayfair*, 615.
17. Anon. (Jan. 1908). The decision of the Royal College of Physicians. *Magazine of the London (Royal Free Hospital) School of Medicine for Women*, (39), 853–855.
18. Mudge, G. P. (1938). The coming of the pioneers and the passing of the veterans. *Magazine of the London (Royal Free Hospital) School of Medicine for Women, 2*(4), 26–33.
19. Roberts, S. (1993). *Sophia Jex-Blake. A woman pioneer in nineteenth century medical reform* (pp. 182–184). Routledge.
20. Todd, M. (1918). *The life of Sophia Jex-Blake* (pp. 447–448). Macmillan & Co.
21. Anon. (1916). Opening of the extension of the school by Her Majesty the Queen. *Magazine of the London (Royal Free Hospital) School of Medicine for Women, 11*(65), 85–95.

# Pioneer Women of the LSMW

# 9

*In Chap. 6, we described the lives and contributions of the remaining members of the so-called Edinburgh Seven. In the founding of the London School of Medicine for Women, some of the wider 'Edinburgh group' were involved as the first cohort of students, together with new members of the 'cause' of women's medical education. This chapter will contain accounts of the lives and contributions of those women not included in Chap. 6. Unfortunately, no information of their later lives of three of them could be found. They were: Jane Russell Rorison, Elizabeth Vinson, and Elizabeth Walker.*

## The First Cohort

It was on 12 October 1874, that the London School of Medicine for Women (LSMW) opened its doors with an initial enrolment of fourteen. These pioneer women students were [1]:

| | |
|---|---|
| Isabel Pryer (Mrs. Thorne) | Elizabeth Vinson |
| Sophia Jex-Blake | Jane Russell Rorison |
| Edith Pechey (Mrs. Pechey-Phipson) | Edith Shove |
| Mary Anderson (Mrs. Marshall) | Elizabeth Walker |
| Alice Ker (Mrs. Ker) | Agnes McLaren |
| Ann Elizabeth Clark | Jane Elizabeth Waterston |
| Isabella Johnstone (Mrs. Foggo) | Fanny Jane Butler |

Twelve of the first cohort had been students at Edinburgh. Of the other two, Fanny Butler had been a medical missionary in Kashmir, while Jane Elizabeth Waterston had been a missionary in South Africa.



M. Rayner-Canham and G. Rayner-Canham, *Pioneers of the London School of Medicine for Women (1874–1947)*, Perspectives on the History of Chemistry,
https://doi.org/10.1007/978-3-030-95439-0_9

They were not an integral class, reflecting their different levels of medical background [2]. To illustrate, they did not take the examinations together: Clark, Foggo, Ker, Vinson, and Shove took examinations in the Winter Session, 1874–75; Clark, Ker, and Anderson (Mrs. Marshall) in the Summer of 1875; and Ker and Rorison in the Winter of 1875–76.

During the first three years, not all these students were attending classes of the LSMW [2]. Those who had transferred from Edinburgh were exempted from repeating courses they had already taken. For example, Pechey worked for a year at the Birmingham and Midlands Hospital for Women, as mentioned in Chap. 5; Thorne spent the second year at the University of Paris; while Ker, at Jex-Blake's suggestion, spent a year in Boston with Lucy Sewall at the New England Hospital for Women.

Not convinced that they would be able to graduate from LSMW, others went to European universities to complete their studies: Jex-Blake, Clark and Pechey went to the University of Bern; Anderson (Mrs. Marshall) went to the University of Paris; McLaren went to the University of Montpelier; and Waterston to the University of Brussels.

## King and Queen's College of Physicians in Ireland

In the early years of the LSMW, as recounted in Chap. 8, the full requirements of a medical degree could not be completed at the School. Though it was possible—and many did—travel to European countries to complete an M.D., such 'foreign' qualifications were not accepted for the practice of medicine in Britain (see Chap. 8). It was the King and Queen's College of Physicians in Ireland (KQCPI) who came to their rescue. Even after the University of London admitted women to examinations, the KQCPI continued to be the most important institution for licencing women from the LSMW through the 1870s and to the end of the 1880s [3]. Among those to whom they granted Licentiate status were: Pechey, and Jex-Blake in 1877; Clark and McLaren in 1878; and Waterson in 1879.

The decision of the KQCPI to first admit women in 1877 was the result of several factors. Dublin had a history of unusual liberality in the education of women. But more important than the societal context, the Council of the KQCPI in the 1870s was composed of senior members of the Irish medical profession who were in favour of the admission of women.

Totally unlike that in Britain, Irish medical education from the 1880s to the 1920s appears to have been egalitarian in nature. Women and men were treated equally in terms of hospital experience, lectures (which they attended together), while prizes and scholarships were open to women on the same terms as men. There was one important exception: there were separate dissecting rooms for male and female students [4].

In 1896, Clara Williams, a student at the KQCPI, wrote to the *LSMW Magazine* of her very positive experiences [5]:

> Nothing in the slightest degree unpleasant has ever occurred, and the professors are unanimous in stating that far from regretting the admission of women to their classes, they consider it has improved the tone of the College considerably. The students are all friendly, there is a healthy spirit of emulation aroused in working together for the various prizes, and an absence of jealousy which augurs well for the future of medical women in Ireland, and reflects favourably on the men as well; we all help each other, and I, for my own part, owe a great deal of my success to the assistance of a few of the senior men students.

## Alice Ker

Alice Jane Shannan Stewart Ker was born on 2 December 1853 at Deskford in Banffshire, Scotland [6]. She was the eldest of the nine children of Margaret Millar Stevenson and Reverend William Turnbull Ker, a Free Church minister. At the age of 18, she moved to Edinburgh to attend 'University Classes for Ladies' her studies including anatomy and physiology. While in Edinburgh, she met Sophia Jex-Blake. Following the rejection of Jex-Blake's petition by the University of Edinburgh, Ker left Edinburgh and moved to London, commencing her studies at the London School of Medicine for Women with the first cohort in 1874. She was an academically gifted student, winning prizes in chemistry, *materia medica*, and practice of medicine. As the School's qualifications were not recognized by the medical register at that time, she sat the examinations of the KQCPI and was awarded her Licentiateship in 1876.

Ker then undertook additional studies in the USA and at the University of Bern. When Ker returned to Britain, she worked as a house surgeon at the Birmingham and Midland Free Hospital for Sick Children, being promoted to Senior Medical Officer in 1881. Then in 1884, she moved to Leeds as a general practitioner, taking over Edith Pechey's hospital post (and her accommodation) when Pechey departed for India [7]. In 1887, Ker returned to Edinburgh working as a self-employed doctor. Taking the Royal College of Surgeons Conjoint Examinations, she was one of only two women in that year to pass the finals.

Ker married her cousin, Edward Stewart Ker, in 1888, and they moved to Birkenhead. There, she opened her own practice, being the only woman doctor in the area. They had two daughters: Margaret Louise and Mary Dunlop while their son died in infancy. Ker's husband died suddenly in 1907, leaving her to raise the two daughters as a single parent. Ker also gave talks and lectures to working-class women in Manchester on topics of sexuality, birth control, and motherhood.

In 1891, the first edition of her book: *Motherhood: A Book for every Woman*, [8] was published. This work, reprinted in 1896, became a classic, as it was as frank and detailed as it was possible to be in Victorian England. An article in the *Glasgow Medical Review* praised it highly [9]: "Medical men would do well to recommend its perusal to the mothers and young wives in their practice, who are quite ignorant of the mysterious laws that govern their bodies, and who would be grateful for information and advice." The review continued by selecting passages from the book [9]:

> Dr. Ker regrets that, "in spite of all that is taught and written on the subject of health, and in spite of everything that is being done for the education of women, the question of periodicity is left entirely out of sight, ..." Further—"Girls must not be allowed to think themselves inferior to boys; ..." "Girls are quite capable of doing quite as much mental work as boys and of doing it equally well,"

Ker was also very active in the women's suffrage movement. In 1916, she moved to London, where she continued School and baby clinics well into her 70s. Ker died in 1943.

## Edith Shove

Edith Shove was born in 1848 in Lewisham, Kent, daughter of John Shove, Corn and Coal Merchant and Mary Hodder Fletcher Cobbett. Shove undertook training in medicine during the early 1870s [10]. For five years, she was an Apothecary Apprentice and Surgeon's Apprentice with Dr. Prior Purvis. Purvis was a strong supporter of women in medicine, as his Obituarist described [11]: "The same liberal bent of mind caused him strongly to support the admission of women into the ranks of the medical profession, and several of the present medical women have frequently expressed their gratitude to him for his efficient help in their early struggles."

Shove was placed first in the Preliminary Apothecaries' Examinations in 1874 [12], and in the same year, joined the first cohort at the London School of Medicine for Women. In 1877, the Senate of the University of London voted that she should be permitted to take the University medical examinations. However, the permission was retracted in response to protests by over 200 male medical graduates.

Subsequently, in 1879, when the University of London finally did open its examinations to women, Shove was one of the first four women to sit the Preliminary Medical Examinations, passing, as the other women in that year did, in the first division. Then in 1881, she gained the Licentiate in Medicine and in Midwifery of the KQCPI, and as a result of her qualification, she was appointed Demonstrator in Anatomy at the LSMW. Graduating with the University of London degree of M.B. in 1882, Shove also carried out joint research on the 'diabetic pancreas' with the French physician Charles Remy, which was published in the same year.

In 1883, Shove was appointed medical officer to the female staff of the British Post Office. This position was the first time that a woman doctor had held a public appointment, and it was bitterly opposed by the medical journal, *The Lancet* [13]:

> The long-expected step of appointing a lady to be Medical Superintendent of the female staff of the General Post Office has been taken. Miss Edith Shove, M.B. Lond. and L.K.Q.C.P.I., is the fortunate lady. Our opinion of this step has been repeatedly expressed, and is not altered by the determination of Mr. Fawcett, to which alone doubtless the appointment must be ascribed. We abundantly recognise the right of ladies to practice on their

> registered diplomas, and as freely admit the ability of the few ladies who have taken medical qualifications. We nevertheless think it a stretch of the power of the Minister to appoint a lady to attend a number of *employées* of her own sex. We do not think the appointment will be agreeable to them.

Shove held this role until at least 1905. She died in London, November 1929, age 82 years.

## Ann Elizabeth Clark

Ann (Annie) Elizabeth Clark was born in Somerset in 1844, daughter of Eleanor and James Clark. Her father founded a shoe manufacturing business [14]. She was educated at a private School in Bath and then lived at home. Clark joined her mother in social causes such as the abolition of slavery and the temperance movement. From the mid-1860s, she was active in the cause of women's suffrage.

In her late 20s, Clark decided upon a career in medicine, first joining Sophia Jex-Blake in Edinburgh, then moving to London to register in the first cohort of the London School of Medicine for Women. At that early date, unable to complete her studies at the LSMW, she transferred to the University of Bern, travelling with Jex-Blake and Edith Pechey, completing her M.D. in 1877. In order to practice in Britain, Clark still needed a British qualification and, as a result, took and passed the Membership Examinations in Medicine and in Midwifery of the KQCPI [15].

After undertaking postgraduate research in Paris, Vienna, and the USA, Clark returned to England in 1878. There, she settled in Birmingham, to take up a position at the Birmingham and Midland Hospital for Women (BMHW). The Hospital had been founded by Robert Lawson Tait as an establishment devoted to the diseases of women, especially gynaecology [16]. Clark continued to work at the BMHW until she retired in 1913, at nearly 70 years of age. Following retirement, she increased her travelling, making frequent journeys to Switzerland and the Tyrol (the northern Italian and western Austrian parts of the Alps). Her cousin, Maida Sturge, had set up a Birmingham Children's Home in the healthy air of the Tyrol, and Clark visited it frequently. Clark died in 1924, age 80.

## LSMW Women Doctors and the Empire

The possibility of a medical career in India was extolled by an article in *Myra's Journal* [17]. The author informed the readers of the 'lady doctorships' of the National Association for Supplying Female Medical Aid to the Women of India, informally known as the 'Countess of Dufferin's Fund' [18]. The article provided the details of the 'doctorship' subsequent to graduation with a medical degree [17]:

> The form of agreement such registered lady doctors connected with the Association will be required to sign, will be to serve for five years in India, and at such stations as the Association may appoint. These ladies have their passages to India provided for them by the Association, which also allows them, on signing the agreement, £10 for outfit and pocket money on the voyage, and a salary of 300 rupees a month from the date of reporting their arrival in India. Beyond this, they will either be provided with free quarters or receive 50 rupees in lieu of such. Each of these lady doctors will, on completion of their service under the agreement, be paid 800 rupees in discharge of all claims, on account of return passage and expenses to England from the stations where they may have been employed.

## Isabella (Isa) Johnstone (Mrs. Foggo)

Isabella (Isa) Margaret Hope Johnstone was born in Calcutta, India, in 1841. Her father, William Johnstone, was a Lieutenant in the 51st Regiment of Bengal Native Infantry, while there is no record of the name of her mother. In October 1861, Isa Johnstone was married in Calcutta to John Thomas Foggo, a Scottish merchant in India [19] and their only child, Jessie Augusta McDonald Foggo, was born in 1864. Her husband died in 1871.

After her husband's death, Foggo travelled from India to Scotland. There, she joined the cohort at the University of Edinburgh in pursuit of a medical education. Of note, she was one of the eleven signatories of the plea for a medical School for women at the University of St. Andrew's (see Chap. 4). As noted at the beginning of this Chapter, Foggo was a member of the first class at the LSMW. She must have completed her medical studies there, for in 1880, she was granted the Licentiate in Medicine and in Midwifery of the KQCPI.

After completion of the medical qualifications, Foggo was appointed House Surgeon at the New Hospital for Women, London [20]. Then in 1887, she returned to India. Foggo took up a position as Manager at the newly opened Lady Dufferin Zenana Hospital in Calcutta. It was reported in the *British Medical Journal* that [20]: "This hospital is intended solely for women. It will afford facilities for medical treatment to the class of purdah women, who were hitherto debarred from going to the female wards of other hospitals."

Lady Dufferin complained that rich Eurasian women should not be availing themselves of Foggo's free services and instead should become private (paying) patients, as Crowther and Dupree describe [21]:

> On a visit to the Calcutta dispensary [in 1888], the Countess found Mrs. Foggo attending to 'several very smart young women in fashionable hats', who claimed they had a 'native' objection to male doctors. 'I consider the doctor already greatly overworked,' wrote Lady Dufferin, ...

In 1889, Johnstone returned to England on sick leave. She died in June 1891 in London.

## Fanny Jane Butler

Fanny Jane Butler was born on 5 October 1850 in Chelsea, London, the daughter of Thomas Butler and Jane Isabella North [22]. Only her brothers were allowed to attend grade School, and they, in turn, taught her at home. At the age of 15, Butler was permitted one year of education at the West London College. In 1872, she travelled to Birmingham to nurse her elder sister. While there, Butler read an article about the need for female medical missionaries in India and this sparked her interest. She was accepted into the India Female Normal School and Instruction Society and passed the Preliminary Examination of the Society with the second-highest grade among a total of 123 students [23].

Later, in 1874, Butler joined the first cohort of students at the London School of Medicine for Women. With the KQCPI opening its examinations to women in 1877, Butler took her examinations there. She was awarded the Licence of Medicine and of Midwifery from the KQCPI in 1880.

Butler travelled to India that same year, staying first in Jabalpur, then travelling to Bhagalpur in 1882. In her $4\frac{1}{2}$ years there, she ran two medical dispensaries and administered to several thousand women patients. After a year's rest back in England, Butler returned to India, this time to Srinagar in Kashmir. Foreigners were not permitted to live in the city, so each day she had to undertake the four-mile journey there and back using a pony or a boat. In the first 7 months, Butler and her staff saw 8832 outpatients and performed 500 operations.

Butler initiated the construction of the first hospital in Srinagar. She had been visited in Srinagar by the British woman explorer, Isabella Bird. Impressed by Butler's work, Bird provided the funds for a hospital to be named the John Bishop Memorial Hospital, after her late husband. Sadly, Butler never saw completion of the Hospital, as she died of dysentery in Srinagar in October 1889.

## Agnes McLaren

Agnes McLaren, like Butler, combined a medical career with missionary service [24]. She was born in Edinburgh on 4 July 1837 [25]. Her father was Duncan McLaren, businessman and Member of Parliament. Her mother died when she was three years old, and it was her father's third wife, Priscilla Bright, who played a major role in her life.

The McLaren family had all been strong supporters of women's education [26]. In particular, her father financially supported the struggles of the 'Edinburgh Seven'; however, he opposed his daughter joining the 'cause'. Nevertheless, Agnes McLaren entered the London School of Medicine for Women as part of the first cohort. Raised a Presbyterian, she shifted to Catholicism, and this caused her to choose the strongly Catholic University of Montpelier in France to complete her medical studies, entering in 1876. In 1877, McLaren was elected to the Governing Body of the LSMW. During this time, she was a visiting physician at the Cannongate Medical Mission Dispensary, Edinburgh. McLaren also opened a clinic in Cannes, France, spending the winters there and the summers at her Edinburgh practice.

In 1898, age 61, McLaren formally converted to Roman Catholicism. Travelling to Rawalpindi, India, with a Catholic mission, she became aware of the inability of Indian women to access medical care by male doctors. Upon her return to England, she established the Medical Mission Committee in London, which financed St. Catherine's Hospital, Rawalpindi. During her search for women to help run the Hospital, McLaren discovered that Catholic Canon Law prohibited Religious Sisters from giving that level of medical care. McLaren decided to remedy this problem, and she petitioned the Pope and Holy See for the removal of the restriction. She retired to Cap d'Antibes, France, where she campaigned against the white slave traffic [27]. McLaren died there in April 1913, before her wish came true with the founding of the Medical Mission Sisters.

## Jane Waterston

Born in 18 January 1843 in Inverness, Scotland, Jane Elizabeth Waterston was the daughter of Agnes Webster and Charles Waterston [28]. Her father was a bank manager. Waterson was educated at Inverness Royal Academy. She decided to become a missionary in the Free Church of Scotland, which was focussed upon South Africa, not India. Waterston travelled to South Africa, being appointed Principal of the Girls' Institution at Lovedale, Cape Colony in 1868.

However, Waterston felt teaching was not her destiny and she travelled back to Britain in 1874, just in time to register as part of the first cohort of students at the London School of Medicine for Women. Like many of the other students, she completed her examinations at the KQCPI. In 1879, she returned to Africa, joining the Livingstonia Mission, Nyasaland. Her intellect and abilities were dismissed by the male missionaries, while she was appalled by their treatment of Africans. After six months, Waterson left and returned to Lovedale, Cape Colony, where she opened a clinic.

In 1883, Waterston moved to the city of Cape Town. Though she was the only woman doctor in the country, she became accepted, in part, because she practiced medicine among women, the poor of Cape Town, and the Xhosa dock workers, whose language she spoke [29]. To further her medical training, in 1887, Waterston travelled to Belgium, being awarded an M.D. from the University of Brussels. Returning to South Africa, during the South African War, she organized relief for the Uitlander women and children who had been deported to the coast from the Boer Republics. When Waterston died in November 1932, her funeral procession was one of the largest seen in Cape Town.

## The Pioneer Women of the LSMW as an 'Invisible College'

The term 'Invisible College' is used to describe a small community of interacting scholars who often met face-to-face, exchanged ideas, and encouraged each other [30]. The pioneering women doctors fit this description, as several continued to interact long after their initial encounters [31]. For example, when Isabel Thorne's last child was born, Matilda Chaplin Ayrton was her medical attendant. And when Sophia Jex-Blake was undergoing an operation, it was Ann Clark who administered the anaesthetic.

Agnes McLaren kept in contact with both Jex-Blake and Edith Pechey [32]. Whenever McLaren returned to Scotland, she visited Jex-Blake in Edinburgh. After Jex-Blake's retirement, McLaren stayed with Jex-Blake in Sussex on her travels between Cannes and Edinburgh. Pechey stopped over in Cannes to visit McLaren whenever Pechey was travelling from India back to England. There were also connections through the Birmingham and Midland Children's Free Hospital and the Birmingham and Midland Hospital for Women between Edith Pechey, Alice Ker, Annie Reay Barker, and Anne Clark [33].

The women also developed links through the militant suffragette movement [31]. Sophy Massingberd-Munday, Jex-Blake's favourite (see Chap. 6), abandoned her medical studies to become a suffragette in London. Alice Ker became the most militant of the group, being an enthusiastic supporter of Emmeline Pankhurst and being sentenced to three months in jail for vandalism. While Edith Pechey joined the movement upon her return from India.

## Mary Bird (Mrs. Scharlieb)

Though not strictly in the first cohort, Mary Ann Dacomb Bird (Mrs. Scharlieb) was an integral part of the story of the early years. Born on 18 June 1845, Mary Bird lived with her grandparents, following the death of her mother [34]. Raised in a strict Evangelical Christian household, she first attended boarding Schools, completing her education at Mrs. Tyndall's School in Brighton. At the age of 19, William Scharlieb, a newly graduated lawyer, proposed marriage to her. Though facing initial family opposition, Bird persisted, and they were married at the end of 1865. William Scharlieb planned to open a practice in Madras, and they sailed to India immediately after the wedding.

In Madras, now as Mrs. Scharlieb, she learned of the lack of provision of medical services for pregnant women. She enrolled at the Madras Medical College in 1875, one of the first four women to attend. Completing the program, in 1878, she was awarded a Licentiate in Medicine, Surgery, and Midwifery. As a result of ill-health, Scharlieb, together with her young children, returned to England on a small ship which four times ran out of the coal fuel during the journey. Back in London, she then enrolled at the London School of Medicine for Women [35]. In 1879, she passed her first examination, and in 1882, she received an M.B. in

**Fig. 9.1** Dr. Mary Scharlieb Lecturing to Students of LSMW, late 1890s. Public domain: London Metropolitan Archives, unknown photographer

Medicine and Surgery, finishing her studies with six weeks in Vienna studying operative midwifery.

On her return to India in 1883, Scharlieb opened a general practice. However, she realized the necessity of a hospital, as she was having to perform operations. To do so, Scharlieb taught her sister how to administer an anaesthetic, and she employed her maid as a nurse assistant. The Indian Government also recognized the need and founded the Royal Victoria Hospital for Caste and Gosha Women, Madras. Scharlieb's typical working day was from 5:30 a.m. until 8 p.m. When an assistant became available, she accepted a position as Lecturer in Midwifery and Gynaecology at the Madras Medical College [36].

Returning again to England, due to recurring ill-health, Scharlieb passed the M.D. examination of the University of London. From 1887 until 1902, she was a Surgeon at the New Hospital for Women. In 1889, Scharlieb was appointed Lecturer of the Diseases of Women at the LSMW (Fig. 9.1). In addition, she was an Examiner for women students at the LSMW. Besides that, Scharlieb also ran a large private practice. She was also a prolific author of books on women, the most acclaimed being: *The Seven Ages of Woman: A Consideration of the Successive Phases of a Woman's Life*, published in 1915 [37]. Scharlieb died in November 1930.

## Commentary

Up to now, the chapters have followed the life stories of individuals who contributed to the founding and early years of the LSMW. Though the founding of the institution had been largely instigated by women, the medical doctors who were to staff the LSMW were almost exclusively male. However, there was a woman-run enclave: the Chemistry Department and the account of this department will be the focus of the next six chapters; the experiences of the students, and an account of the lives and contributions of the women faculty and staff.

## References

1. Thorne, I. (1905). *Sketch of the foundation and development of the London School of Medicine for Women* (p. 19). G. Sharrow.
2. McIntyre, N. (2014). *How British women became doctors: The story of the Royal Free Hospital and its Medical School* (pp. 53–54). Wenrowave Press.
3. Kelly, L. (Feb. 2013). 'The turning point in the whole struggle': The admission of women to the King and Queen's College of Physicians in Ireland. *Women's History Review, 22*(1), 97–125.
4. Kelly, L. (2010). 'Fascinating Scalpel-wielders and Fair Dissectors': Women's Experience of Irish Medical Education, c. 1880s–1920s. *Medical History, 54*(4), 495–516.
5. Williams, C. L. (Jan. 1896). A short account of the school of medicine for men and women, RCSI. *Magazine of the London (Royal Free Hospital) School of Medicine for Women,* (3), 91–132.
6. Cowman, K. (23 Sept. 2004). Ker, Alice Jane Shannan Stewart (1853–1943), doctor and suffragette. *Oxford Dictionary of National Biography*. https://doi-org.qe2a-proxy.mun.ca/https://doi.org/10.1093/ref:odnb/63874. Accessed 28 Aug. 2021.
7. Crowther, M. A., & Dupree, M. W. (2007). *Medical lives in the age of surgical revolution* (p. 161). Cambridge University Press.
8. Ker, A. (1891). *Motherhood: A Book for every woman*. John Heywood.
9. Anon. (1892). Reviews. *Glasgow Medical Journal, 37*, 224–225.
10. Scharlieb, M. (14 Dec. 1929). Obituary: Edith Shove M.D. Lond. *British Medical Journal, 2*, 1137.
11. Anon. (13 Jun. 1908). Obituary: Prior Purvis, M.D. Lond. *British Medical Journal, 1*, 1463.
12. Anon. (30 Jul. 2021). Edith Shove. https://en.wikipedia.org/wiki/Edith_Shove. Accessed 28 Aug. 2021.
13. Anon. (17 Mar. 1883). The Appointment of Miss Shove MB to the Post Office. *The Lancet, 121*(3107), 468.
14. Gil, S. https://stumblingstepping.blogspot.com/2015/02/quaker-alphabet-blog-2015-c-for-annie.html. Accessed 4 Oct. 2021. (Blog).
15. Anon. (29 May 2021). Annie Clark (physician) https://en.wikipedia.org/wiki/Annie_Clark_(physician). Accessed 24 Sept. 2021.
16. Golditch, I. M. (2002). Lawson Tait: The forgotten gynaecologist. *Obstetrics and Gynecology, 99*(10), 152–156.
17. "J.N.P." (Mar. 1889). Employments for women: Myra's Monthly Guide to women desirous of earning a livelihood or adding to an income. Lady doctors for India. *Myra's Journal*, 126–127.
18. Lal, M. (1994). The politics of gender and medicine in colonial India: The Countess of Dufferin's Fund. *Bulletin of the History of Medicine, 68*, 29–66.
19. Ref. 7, Crowther & Dupree, p. 162.
20. Anon. (1887). India and the Colonies: Calcutta. *British Medical Journal, 2*, 439.
21. Ref. 7, Crowther & Dupree, p. 315.
22. Forbes, G. (25 May 2006). Butler, Fanny Jane (1850–1889), physician and medical missionary. *Oxford Dictionary of National Biography*. https://doi.org/10.1093/ref:odnb/61066. Accessed on 22 Sept. 2021.
23. Tonge, E. M. (1930). *Fanny Jane Butler, pioneer Medical Missionary*. Church of England Zenana Missionary Society.
24. Burton, K. (1946). *According to the pattern: The story of Dr. Agnes McLaren and the Society of Catholic Medical Missionaries*. Longmans, Green and Co.
25. Anon. (23 Sept. 2021). Agnes McLaren. https://en.wikipedia.org/wiki/Agnes_McLaren. Accessed 23 Sept. 2021.
26. Ref. 7, Crowther & Dupree, p. 38.
27. Anon. (26 Apr. 1913). [obituary]: Dr. Agnes McLaren. *British Medical Journal, 1*, 917.
28. Van Heyningen, E. (25 May 2006). Waterston, Jane Elizabeth (1843–1932). *Oxford Dictionary of National Biography*. https://doi.org/10.1093/ref:odnb/59011. Accessed 22 Sept. 2021.

29. Van Heyningen, E. (1996). Jane Elizabeth Waterston–Southern Africa's first woman doctor. *Journal of Medical Biography, 4*(4), 208–213.
30. Anon. (9 Sept. 2021). Invisible College. https://en.wikipedia.org/wiki/Invisible_College. Accessed 23 Sept. 2021.
31. Ref. 7, Crowther & Dupree, p. 170.
32. Ref. 7, Crowther & Dupree, p. 166.
33. Ref. 7, Crowther & Dupree, p. 157.
34. Jones, G. (23 Sept. 2004). Scharlieb [née Bird], Dame Mary Ann Dacomb (1845–1930). *Oxford Dictionary of National Biography*. https://doi.org/10.1093/ref:odnb/35968. Accessed 13 Oct. 2021.
35. Anon. (29 Nov. 1930). Obituary: Dame Mary Scharlieb, D.B.E., LL.D., M.D., M.S. *British Medical Journal, 2*, 935–937.
36. Collinson, S. R. (1999). Mary Ann Dacomb Scharlieb: A medical life from Madras to Harley Street. *Journal of Medical Biography, 7*(1), 25–31.
37. Scharlieb, M. (1915). *The seven ages of woman: A consideration of the successive phases of a woman's life*. Cassell & Co.

# Chemistry at the LSMW 10

*As shown in the London School of Medicine curriculum in Chap. 8, chemistry was a crucial part of the studies from the School's beginning in 1874. It is the Chemistry Department, almost exclusively run by women chemists, which provides a thread in this book, through the intervening years between the School's foundation to its final absorption and disappearance. An added reason for choosing this context is that, in the pages of the London School of Medicine Magazine, the women students expressed their feelings with the greatest clarity. Enthusiasm, fear, humour: All could be found in the students' comments about the chemistry courses and their women instructors. Here, we provide a glimpse into this part of a students' education through the decades at the School.*

## LSMW Chemistry Facilities

In her history of the early years of the London School of Medicine for Women (LSMW), Isabel Thorne recalled [1]: "… the Lectures on Chemistry were given in a room on the left-hand side of the garden entrance to the old building, which was also used for Practical Chemistry in the summer, …" (Fig. 10.1). It was not until 1898 that dedicated spaces for the sciences, including laboratories, were completed. The journal, *The Lancet*, reported on the new facilities [2]:

> The laboratories at the London School of Medicine for Women in Handel-street, Brunswick-square, which have been recently built, form a large block of four storeys and comprise a dissecting room and laboratories for physiology, chemistry, and physics, with accessory rooms. ... The chemical laboratory is not yet completely finished; it will contain tables for about 70 students, and an adjoining room is fitted up as a private laboratory for the lecturer.

The formal opening of the laboratories was described in detail in the *Magazine of the London (Royal Free Hospital) School of Medicine for Women* (henceforth



M. Rayner-Canham and G. Rayner-Canham, *Pioneers of the London School of Medicine for Women (1874–1947)*, Perspectives on the History of Chemistry,
https://doi.org/10.1007/978-3-030-95439-0_10

**Fig. 10.1** Students in the old LSMW Chemistry Laboratory, 1894. Public domain: London Metropolitan Archives, unknown photographer

called the *LSMW Magazine*). The ceremony, held in the new LSMW Chemistry Laboratory, showed strong royal support for a women's medical School. The guests of honour were Edward, Prince of Wales; Alexandra, Princess of Wales; and the Bishop of London. It was Princess Alexandra (not the Prince) who opened the facilities, a notable departure from precedent, as was remarked at the time [3]: "Then the Princess herself declared the Schools to be open, an act for her unusual, but on that account particularly gracious on this occasion, as coming from a woman to women."

In a subsequent issue of the *LSMW Magazine*, students added their own comment [4]:

> ... the new laboratories have proved a great improvement on the old. They are lighted by electricity throughout. ... The Chemistry Laboratory is greatly appreciated. Its fittings are very complete, and it, like all the rooms in the new building, is a most pleasant room in which to work.

As discussed in Chap. 8, there was an additional expansion of the LSMW in 1916, necessitated by a considerable increase in student numbers. This extension included a new LSMW Organic Chemistry Laboratory which was reported in the *LSMW Magazine* [5]: "In the 'Maude du Cros' Organic [Chemistry] Laboratory Her Majesty was received by Sir Arthur and Lady du Cros, Mr. J. A. Gardner, M.A., F.I.C., Lecturer in Organic Chemistry and Miss S. T. Widdows, B.Sc., Lecturer in Organic Chemistry." The fact that the laboratory had been named the Maude du Cros laboratory, after Lady Maude du Cros (née Gooding) would suggest that Sir Arthur du Cros, known for his generous donations to good causes, had part-funded the new organic chemistry laboratory [6].

## LSMW Student Writings on Chemistry

In the early decades, practical chemistry was a periodic topic of the students' literary submissions to the *LSMW Magazine*. As an example, in 1912, I.N. Glough authored a fictitious submission by 'Vanitas' to an imaginary advice columnist [7]:

> Vanitas writes: "What sort of a chemistry overall should I choose?"
> We fear our advice will come a little late, Vanitas, but perhaps you will have worn out an apron or two by the time this article appears. The great thing is to have a background for the various chemical dyes you are certain to collect on your overall. Red and blue go splendidly with $Am_2S$, HCl, $HNO_3$ and the blackness of London dust.
> Have you seen the new reversible pinafores. They are supplied by Messrs. Peace Bros., and we only wish they had been in vogue in our chemistry days. On one side they are of various colours, blue, green, pink, according to taste. On the other they are printed to look just like your chemistry bench. The more expensive ones have even taps and apparatus to match. So when the Demonstrator is at the same time on the warpath and on your track all you have to do is to reverse your pinafore and—where are you? Invisible till the storm blows over!

In terms of staffing the chemistry laboratories, there were the Demonstrators and under them, the Lab-Boys. In an issue of the *LSMW Magazine* of 1911, the author points out that it was the Lab-Boy who was of prime importance in the chemistry laboratory [8]:

> We sometimes wonder who it is who really is the all-presiding genius of the laboratory, is it the demonstrator, the student, or is it Knowledge with a big *K*?
> No, it is the "lab. boy."
> It is he who teaches us, is patient with us, leads us like the children that we are, but above all it is he who cheers us on our way and talks to us of people and of things.
> The chief characteristic about the chemical laboratory boy is his dirt, a property which he shares in common with all born chemists. Unlike the others [physiology and bacteriology] he makes great efforts to get at the theory of his work, attends the Polytechnic at night and demands days off for his "Matric"; thus he is better equipped than his colleagues of the other sciences and his influence tends to help the serious studies of those he waits on, rather than to develop their character by enlarging their acquaintance with life in general.

## LSMW Student Writings on Chemical Accidents

Safety seems to have been of little concern, as this commentary in the Summer Term, 1895, issue of the *LSMW Magazine* illustrates [9]:

> This is the term for practical histology and chemistry, studies in which our zeal is evidenced by the crimson and orange stains with which we adorn our fingers, and the red-edged rents which splashes of sulphuric acid make in our gowns. Practical chemistry is not without its excitements, as some of us can bear witness; for have we not had several fine explosions this term in the laboratory, resulting in the mutilation of more than one of our number?

According to a commentary in an issue of the *LSMW Magazine* of 1901, explosions still seem to be an expected part of practical chemistry [10]:

> The chemistry lectures came to an end last term, but practical work still continues twice a week, and twice a week there are hair-breadth escapes from death by poisoning, owing to the efficacy of the draught cupboards, not to speak of the exemplary caution of the students in the matter of "Unknowns." The few explosions that occurred last term, although alarming in report, were fortunately not attended by any fatal results.

In a 1912 issue of the *LSMW Magazine*, the chemistry laboratory inspired fear and foreboding in this medical student [11]:

> In practical chemistry the elements themselves do conspire against me, and a mixture that on my right and on my left behaveth with seemliness doth explode and knock me over. The noisesome fumes that brood over the place do somewhat concentrate themselves in my nostrils, and had I the world's rheumatism in my limbs assuredly it had been cured by all the $H_2S$ that I have inhaled.

In addition to the regular chemistry laboratory work, a practical examination was a requirement for the chemistry courses. The student who wrote this account in the *LSMW Magazine* in 1920 was clearly 'having a bad day' [12].

> We were doing practical Phys. [Chem.] that morning, of the variety "chemical estimations." My partner and I bestowed the utmost care upon our Kjeldahl distillation [13]: we cherished its receiver with cold baths, and its splash-head with wet dusters: I need perhaps hardly mention what occurred—(possibly you know that feeling too?). But I still think the staff might have been more sparing with their remarks concerning the expense and rarity of Erlenmeyer flasks, the impossibility of replacing splash-heads, and the removal of all varnish from the bench: after all, *we* had to pay, and most of the soda was on our garments, and moreover we had to begin another estimation. There were other minor casualties that morning in addition to the main catastrophe, but I will spare my readers.

## LSMW Student Inorganic Chemistry Analysis

Inorganic Chemical Analysis was a significant component of practical chemistry. 'Qual', as it was commonly known, is a means of qualitatively identifying metal ions in a solution [14]. Categories are used to sequentially identify ions, for example, whether the metal ions form an insoluble chloride compound. Following categorization (IA, IIA, IIB, and so on) specific tests are then used to identify individual metal ions. This recorded conversation in a 1912 issue of the *LSMW Magazine* provides one such mention [15]:

> FIRST YEAR STUDENT (*doing chemical analysis*)
> *Reads*: "Tests for the carbonates (salts of hypothetical acid $H_2CO_3$)."
> *Much worried, asks Demonstrator*: "Please, Arthur [stores boy] hasn't got hypothetical acid, or any of its salts; what *am* I to do?"
> [For the reader's information, the 'hypothetical acid $H_2CO_3$' is carbonic acid, the 'parent' acid of carbonates. However, carbonic acid does not exist, as such; dissolving carbon dioxide in water, instead, it simply gives $CO_2(aq)$. Hence the reference to 'hypothetical' acid].

We have shown elsewhere that women students, in a wide range of institutions, framed their chemical experiences in poetic verse [16]. The chemistry students of the LSMW were no exception. For a 1916 issue of the *LSMW Magazine*, one student composed the following lengthy and beautifully crafted mnemonic on qualitative inorganic analysis. Here, only the verses for 'Group IA' and 'Group VI' are included [17]:

**A Mnemonic of Inorganic Analysis**
(Brackets indicate confirmatory tests)
If you learn this little rhyme,
You'll pass First medical-in time!
Given a salt you do not know
Start analysing, as below.

### Group IA

Add HCl, and then you'll get
The metals of Group I., you bet.
Add water, and you'll find the lead
Without a word, has softly fled.
(If on this point your partner wrangles
KI will give you golden spangles.)
You then proceed to add ammonia,
The silver, you perceive, has flown-i-a!
The blackened mercury will lead yer
(To add to this some aqua regia
If stannous chloride's added here
On warming, Hg will appear.)

### Group VI

Heat orig. sol. with sodium hy.
Ammonia hits you in the eye.
Evap. to dryness. By much toil
Save up and buy a plat'num foil.
And heat on this the residue,
Add HCl, divide in two
Add Chlo. and hy. of $NH_3$
Then sodium phosphate shows Mg.
(With charcoal test Mg is pink)
Na and K you know, I think.
And thus, dear friends you should be able
To analyse without a table.

## LSMW Student Practical Organic Chemistry

The preparation of organic compounds was a significant part of organic chemistry practical work. At the 50th Jubilee celebrations of the LSMW, a student, touring the facilities, mused in a 1924 issue of the *LSMW Magazine* [18]:

> When we felt more intellectually minded, we could go and gasp in wonderment at the crystals in the Organic [Chemistry] Laboratory, remembering the little, mucky messes, which were our finest efforts—or we could admire the scientific experiments in the Physiology and Chemistry Laboratories.

In a 1939 issue of the *LSMW Magazine*, a LSMW student put her thoughts on practical organic chemistry in poetic verse [19].

### Thoughts on Organic Chemistry

Awful stinks and stenches,
Cleaning up of benches.
("Don't put the burner on, the ether'll catch alight.").
Complicated formulæ,
Acids, salts and alkali,
Working out of problems that take you half the night.
Biot, Bel and Pasteur,
They could always master.
The problem of a carbon floating round in space.
But my weak intelligence.
Doesn't seem to see the sense.
Of a tetrahedral form with an asymmetric face.
I wish I needn't do it.
But if I don't I'll rue it.
And still I think I'd really rather have a try.
So here's to all Organic.
(I'm in an awful panic).
Oh, Fischer—Biot—Pasteur, help me, or I die!

## Commentary

In this chapter, we have provided a sense of the women's students attitudes to the chemistry courses which were a prerequisite for a medical degree. What comes through most strongly is the enthusiasm of the students—even at the expense of safety—often expressed in verse. Their comments also illustrated the breadth and depth of the chemistry studies required for the medical degree programme at LSMW. The next five chapters will focus upon the women chemistry staff who taught the chemistry courses and supervised the chemistry laboratories at the LSMW.

## References


1. Thorne, I. (1905). *Sketch of the foundation and development of the London School of Medicine for Women* (p. 18). G. Sharrow.
2. Anon. (16 Jul. 1898). New laboratories at the London School of Medicine for Women. *The Lancet, 152*, 160–161.
3. Anon. (Oct. 1898). Opening ceremony of the new block at the school, July 11th, 1898. *Magazine of the London (Royal Free Hospital) School of Medicine for Women*, (11), 453–456.
4. Anon. (Jan. 1899). School news. *Magazine of the London (Royal Free Hospital) School of Medicine for Women*, (12), 506.
5. Anon. (1916). Opening of the extension of the school by Her Majesty the Queen. *Magazine of the London (Royal Free Hospital) School of Medicine for Women, 11*(65), 85–95.
6. Lunney, L. (Oct. 2009). Du Cros, Sir Arthur Phillip. In J. McGuire, & J. Quinn (Eds)., *Dictionary of Irish Biography*. Cambridge University Press. https://www.dib.ie/biography/du-cros-sir-arthur-philip-a2789. Accessed 15 Oct. 2021.
7. Glough, I. N. (1912). Notes and queries. *Magazine of the London (Royal Free Hospital) School of Medicine for Women, 8*(53), 125.
8. Anon. (1911). 'The lab. boy.': An appreciation. *Magazine of the London (Royal Free Hospital) School of Medicine for Women, 7*(48), 417–418.
9. Anon. (Oct. 1895). School news. *Magazine of the London (Royal Free Hospital) School of Medicine for Women*, (2), 77.
10. Anon. (May 1901). Hospital and school news. *Magazine of the London (Royal Free Hospital) School of Medicine for Women*, (19), 804.
11. Anon. (1912). A medical student's purgatory. *Magazine of the London (Royal Free Hospital) School of Medicine for Women, 8*(52), 70.
12. Anon. (1920). The lament of a March examinee. *Magazine of the London (Royal Free Hospital) School of Medicine for Women, 15*(76), 98–99.
13. Anon. (3 Feb. 2021). Kjeldahl method. https://en.wikipedia.org/wiki/Kjeldahl_method. Accessed 4 Feb. 2021.
14. Svelha, G. (1996). *Vogel's qualitative analysis* (7th ed.). Pearson Publishing.
15. Anon. (1912). Heard about hospital. *Magazine of the London (Royal Free Hospital) School of Medicine for Women, 8*(51), 44.
16. Rayner-Canham, M. F., & Rayner-Canham, G. W. (2011). British women, chemistry, and poetry: Some contextual examples from the 1870s to the 1940s. *Journal of Chemical Education, 88*, 726–730
17. 'EDF'. (1916). A mnemonic of inorganic analysis. *Magazine of the London (Royal Free Hospital) School of Medicine for Women, 11*(64), 71–74.
18. 'A Present Student'. (1924). The Jubilee. *Magazine of the London (Royal Free Hospital) School of Medicine for Women, 19*(89), 125–127.
19. Anon. (Mar. 1939). Thoughts on organic chemistry. *Magazine of the London (Royal Free Hospital) School of Medicine for Women, 2* (new series), 120.

# Lucy Everest Boole 11

*When the London School of Medicine opened in 1874, chemistry was taught by Charles Heaton. The first woman to hold a position in the Chemistry Department of the School was Lucy Everest Boole who was appointed in 1891. This Chapter will essentially be her story, but first, we start with a brief account of Heaton and his role.*

## Charles Heaton

Charles William Heaton, born 1835, had commenced a study of practical chemistry as a student at St. Bartholomew's Hospital Laboratory [1]. Despite lacking any formal chemistry qualification, Heaton's aptitude for chemistry resulted in him being appointed chemical teacher at the Royal Benevolent College, Epsom. By 1861, he was a Demonstrator of Practical Chemistry at Charing Cross Hospital. While at the Hospital, he wrote a book on practical chemistry: *The Threshold of Chemistry: An Experimental Introduction to the Science* [2].

In 1862, Heaton was promoted to Lecturer in Chemistry at the Charing Cross Hospital Medical School, a position which he held up to the time of his death. His continued interest in practical chemistry teaching was shown by his translation from the German of a handbook on experimental chemistry in 1872 [3]. Concurrent with his position at Charing Cross Hospital, on the opening of London School of Medicine for Women (LSMW) in 1874, Heaton was appointed Lecturer in Chemistry at the School [4].

Heaton did not teach organic chemistry, a requirement for the Preliminary Scientific Examination. It is of note that Heaton's own book [2] did not include any organic chemistry, and it may indicate his ignorance of the field. Mary Scharlieb (see Chap. 9), a student at that time, commented in her reminiscences published in the *LSMW Magazine*, how the students overcame this hurdle [5]:



M. Rayner-Canham and G. Rayner-Canham, *Pioneers of the London School of Medicine for Women (1874–1947)*, Perspectives on the History of Chemistry,
https://doi.org/10.1007/978-3-030-95439-0_11

> At that time [1879] it was by no means easy to obtain the necessary instruction. Mr. Charles Heaton, the lecturer on Chemistry, did not include Organic Chemistry in the syllabus of his lectures, and those students who were preparing for the examination in question had to seek instruction elsewhere. This we found at the Pharmaceutical Society in Bloomsbury Square. We were made most kindly welcome and we had very good work, chiefly under the guidance of Professor Atwood.

In later years, Heaton was plagued by ill-health. He resigned in 1893 and died in the same year.

## Boole's Family History

The Booles were an extremely talented family [6]. George Boole, her father, was a renown Professor of Mathematics at Queen's College, Cork, Ireland [7]. Mary Everest, her mother, was taught in France by tutors, then she attended an English boarding School. It was not expected that she would acquire an occupation. Nevertheless, Everest was an avid reader, particularly works on religion and the ancient occult sciences, while algebra and arithmetic were her special interests [8]. Everest and Boole had five daughters: Lucy Everest Boole, the fourth daughter being born in Cork, Ireland, on 5 August 1862 [9].

Following her husband's death in 1864 at age 49, Everest saw her future as that of following from her deceased husband's work. She was a prolific author over an incredible range of fields, including mathematics education [10] and science education [11].

In 1865, Everest, with the children, moved to London. She contacted prominent Christian Socialist and educationalist, Frederick Denison Maurice, who had founded Queen's College, Harley Street, the first academic College for girls in England [12]. Maurice offered her a position of librarian and warden of a student residence. Everest's biographer, Mary Creese, commented [8]: "By 1873 Mary [Everest] Boole had come to be regarded by the College administration as somewhat unstable and had had to leave."

Life for the Boole daughters was grim. H. S. M. Coxeter, biographer of Alicia Boole, one of Lucy Boole's sisters, wrote that Everest and the five daughters lived in [13]:

> ... a poor, dark, dirty, uncomfortable lodging in Marylebone. At first all five girls slept in one sunless and dismal bedroom. No privacy was possible; even cleanliness was difficult. They had no education; everyone's nerves were on edge; and Mrs. Boole's friendship with James Hinton ... brought into the house a continual stream of cranks. In their one sitting room, they talked endlessly about subjects that Alice, Lucy, and Ethel were too young to understand but not too young to brood over.

Everest's friend, James Hinton, was a gifted aural surgeon who became notorious for his philosophical writings, including advocacy and practice of polygamy [14]. Things improved intellectually for Lucy Boole when she was about fifteen years of age by the arrival of Hinton's son, Charles Howard Hinton. Charles Hinton was a recently graduated mathematician and writer of science fiction works (who her sister, Mary Ellen, married in 1880) [15]. As Coxeter added [13]:

> He brought a lot of little wooden cubes and piled them up into shapes in his attempt to elucidate the four-dimensional hypercube, or *tesseract*. He set the three youngest girls [Alice, Lucy, and Ethel] the task of memorizing the arbitrary list of Latin words (*Decus*, *Pulvis*, etc.) by which he had named the little cubes. Lucy, being a child with a strong sense of duty, worked hard. ...

## Boole's Pharmaceutical Career

Lucy Boole's career started with an apprenticeship in pharmacy with Isabella Skinner Clarke–Keer (see Chap. 2) [16]. Clarke, the first woman to be registered as a Pharmaceutical Chemist, had opened her own pharmacy on Spring Street, Paddington, in 1876 [17]. It is unknown how Boole learned of the pharmacy, but the London suburbs of Paddington and Marylebone (the location of the Boole residence) were very close. Presumably, Boole, scientifically inclined but without formal education, saw an apprenticeship as a rare career opportunity for herself. The paths of Boole and Clarke were to cross again at the London School of Medicine for Women, as Clarke was later appointed as Tutor in Pharmacy and taught a Practical Pharmacy course there.

While apprenticing at the pharmacy, Boole studied part-time at the Pharmaceutical Society's School of Pharmacy under Theophilius Redwood, Professor of Chemistry and Practical Pharmacy; John Attfield, Professor of Practical Chemistry; and Wyndham Rowland Dunstan; Professor of Chemistry [18]. Attfield had established himself as a proponent for women. In an account of early British women researchers in pharmacy, Joe Shellard had noted [19]:

> In July 1874 Professor Atfield sought permission from the Council [of the Pharmaceutical Society] to admit ladies to his practical classes but this was refused and it was several years later before ladies were allowed to attend practical classes at the Society's School.

Boole passed the Minor in October 1887 and the Major in April 1888 [20].

In 1887, in addition to holding a Professorship in Chemistry, Dunstan had been appointed Director of the new Research Laboratories of the Pharmaceutical Society [21]. Immediately upon her graduation, Dunstan offered Boole the position of his Research Assistant. As such, she was the first woman researcher in the field of pharmacy. Dunstan, too, must have been sympathetic to the cause of women's education as he was vice-President of the Girls' Public Day School Trust. The Trust controlled a network of English academic day Schools for girls which had a strong emphasis on chemistry in their curricula [22].

Dunstan's early work was concerned almost wholly with devising methodologies for the assay of active components in the natural drugs of the *British Pharmacopoeia*. He assigned Boole the task of researching the analysis method for Tartar Emetic, chemical name, potassium antimonyl tartrate, formula $K_2Sb_2(C_4H_2O_6)_2$.

This toxic compound was a traditional and, at the time, commonly used method of inducing vomiting [23]. In addition, in the late 1800s and early 1900s, it was used as a remedy for alcohol intoxication.

By means of experiments, Boole found that the 'official' method of analysis in the *British Pharmacopoeia* using a gravimetric procedure was severely flawed. She devised a new method by titrimetric analysis. Boole and Dunstan presented a paper to an evening meeting of the Pharmaceutical Society on 14 November 1888, and the research was published the following year [24]. According to Shellard [19]: "... the procedure suggested by Lucy Boole was included in the 1898 *British Pharmacopoeia* and remained the official method of assay until 1963 when the sodium bicarbonate was replaced by borax [though Boole's methodology stayed the same]."

## Boole at the LSMW

In 1891, Lucy Boole was appointed Demonstrator in Chemistry at the LSMW under Heaton [19]. It was noted in the LSMW Report for 1892 that [25]: "The teaching of Chemistry has been strengthened by the appointment of Miss Lucy Boole, who holds a tutorial class, which is found most useful." Shortly after, because of Heaton's ill-health, she took over his duties. Upon his resignation in 1893, Boole was promoted to Lecturer in Chemistry as his replacement [26]: "Miss Lucy Boole becomes Lecturer in Chemistry. Miss Forrest B.Sc. (London) [Elsie Forrest, B.Sc., 1892, see Chap. 15] was recommended to succeed Miss Boole as Demonstrator of Chemistry."

While holding her teaching positions, Boole continued undertaking research work with Dunstan at the Research Laboratory of the Pharmaceutical Society. Her research was on croton oil, an oil prepared from the seeds of a tropical tree [27]. The oil caused inflammation and blisters when applied to the skin, together with many other injurious effects. In the opening paragraph of their subsequent publication, Dunstan and Boole describe the reason for their research [28]:

> The nature of the vesicating, or more strictly the pustule-producing, constituent of croton oil is a long-outstanding problem in chemical pharmacology. Many attempts have been made to disentangle from the complicated mixture of fatty acids and glycerides expressed from the seeds of *Croton tiglium*, known as croton oil, a single substance exhibiting the remarkable power of raising pustules on the skin.

At the end of the 13-page paper, they admit defeat, while warning of the severe health-hazards of working with 'croton resin' the active component [28]:

> The task of unravelling the constitution of this substance would be a very difficult one if not an impossible one with our present knowledge, and the difficulty is greatly increased by the circumstance that it is apparently incapable of being crystallized. The amount of croton resin present in croton oil is small; probably it does not exceed 3 per cent. The extraction of the resin is not only a very tedious and troublesome operation, but it is one not unattended with danger to health. It attacks the mucous membrane and the skin with great severity;

> and we have evidence that when the body is continually exposed to its action on one part or another—by contact with the hand, or by the access of minute particles to the nose and throat, as may occur during manipulation of the dry substance, or during the evaporation of solutions containing it—the resin produces a series of secondary effects which seem worthy of special study.

Towards the end of 1897, Boole had submitted her resignation from the LSMW on grounds of deteriorating health [19]. Upon reading the account of the effects of 'croton resin' in her research, it seems extremely likely that the health deterioration was related to her contact with croton resin. In fact, in more recent research, eleven irritant and tumour-promoting compounds were isolated from 'croton resin' [29].

Wishing to keep her, the Council of the LSMW divided her duties, assigning her as Teacher of Practical Chemistry while hiring Clare de Brereton Evans (see Chap. 12) to be Lecturer in Chemistry. The news was noted in the *LSMW Annual Report* [30]: "Miss Boole, F.I.C., having resigned the Lectureship in Chemistry, Miss Evans, D.Sc., has been appointed to lecture on Chemistry during the Winter Session, 1897–98, Miss Boole retaining the direction of the Practical Classes."

## Professional Activities

The Institute of Chemistry had accidentally admitted its first woman, Emily Lloyd, as an Associate in 1892 [31]. Having admitted a woman as an Associate, they could not bar women from the more prestigious ranks. And it was Lucy Boole who, in 1894, was elected as the first woman Fellow of the Institute [32].

The other organization for British chemists, the Chemical Society, since its founding in 1841, had been steadfast in its opposition to the admission of women chemists. In 1904, 19 women chemists signed a petition to the Chemical Society stating [31]: "We, the undersigned, representing women engaged in chemical work in this country desire to lay before you an appeal for the admission of women to Fellowship in the Chemical Society." One of the 19 signatories was Lucy Boole. In a response to the petition, the Chemical Society Council voted unanimously for a proposal to change the bye-laws so as to admit women. Sadly, in a required vote among the over 2700 Members of the Society, only 45 attended the Extraordinary Meeting, of whom 23 voted against admission of women. Their cause was lost.

## Later Years

Boole continued in the position as Teacher of Practical Chemistry for another seven years, dying in December 1904 at age 42. Her Obituarist in the *LSMW Magazine* commented [33]:

> In spite of much physical weakness Miss Boole during these years has only twice been absent from her work for any lengthened period. She has had the fulfilment of her wish, for her a very genuine one, that life might not long outlast power of work, for she only left

us one term before her life ended. ... It was certainly happier for her, for work to her was life, and its absence merely existence.

Included in the Obituary was an appreciation from one of her former students [33]:

> Miss Boole was no believer in 'cram-work,' it was the real deeper meaning of her science that she cared about; and while she taught us with conscientious care the facts necessary for us to know for our examinations, those who knew her well realised that to her that part of the subject was only the threshold to an inner world of knowledge untouched by examination requirements.

Late in 1905, donations were made to purchase an unusual memorial to Boole as described in the *LSMW Magazine* [34]:

> The School Staff, and those among the present students, both at School and Hospital, who knew Miss Boole, have subscribed for a small memorial to her. A Freezing Point Apparatus has been obtained from Messrs. Gallenkamp & Co. It has a copper trough on which is the inscription: "Presented to the Chemical Laboratory in memory of Lucy Everest Boole, F.I.C., Lecturer and Demonstrator in Chemistry, Oct. 1891-Dec.,1904." It is hoped this will be a permanent and suitable, if not very prominent form of memorial. It will certainly be of practical use to the students, which is, we believe, what Miss Boole herself would have wished.

## Commentary

It is sad to contemplate what Boole might have accomplished had she not, as seems most likely, died prematurely as a result of exposure to croton oil. Fortunately, there were a series of talented women chemists to take her place over the ensuing decades, the first of which was Clare de Brereton Evans, the subject of the next Chapter.

## References

1. "S.A.V." (1894). Obituary notices: Charles William Heaton. *Journal of the Chemical Society, Transactions*, 386–388.
2. Heaton, C. W. (1861). *The threshold of chemistry: An experimental introduction to the science*. Chapman & Hall.
3. Heaton, C. W. (1872). *Experimental Chemistry, founded on the work of Dr. Julius Adolph Stöckhardt: A handbook for the study of the science by simple experiments*. Bell and Daldy.
4. Thorne, I. (1905). *Sketch of the foundation and development of the London School of Medicine for Women* (p. 18). G. Sharrow.
5. Scharlieb, (Mrs.) C.B.E., M.D., M.S. (1923). The early days of our school. *Magazine of the London (Royal Free Hospital) School of Medicine for Women, 19*(85), 89–90.
6. Chas, M. (2019). The extraordinary case of the Boole family. *Notices of the American Mathematical Society, 66*(11), 1853–1866.
7. MacHale, D. (1985). *George Boole: His life and his work*. Boole Press.
8. Creese, M. R. S. (23 Sept. 2004). Boole [*née* Everest] Mary. *Oxford Dictionary of National Biography*. https://doi.org/10.1093/ref:odnb/38817. Accessed 7 Jan. 2021.

9. Anon. (26 May 2021). Lucy Everest Boole. https://en.wikipedia.org/wiki/Lucy_Everest_Boole. Accessed 12 Oct. 2021.
10. Tahta, D. G. (Ed.). (1972). *A Boolean anthology: Selected writings of Mary Boole on mathematical education*. Association of Teachers of Mathematics.
11. Boole, M. E. (1904). *The preparation of the child for science*. Clarendon Press, Oxford.
12. Tweedie, E. (Ed.). (1898). *The first college open to women, Queen's College London: Memories and records of work done, 1848–1898*. Queen's College.
13. Coxeter, H. S. M. (1987). Alicia Boole Scott (1860–1940). In L. S. Grinstein & P. J. Campbell. (Eds.), *Women of mathematics: A Biobibliographic sourcebook* (pp. 220–224). Greenwood Press.
14. Anon. (16 Sept. 2021). James Hinton (surgeon). https://en.wikipedia.org/wiki/James_Hinton_(surgeon). Accessed 12 Oct. 2021.
15. Anon. (20 July 2021). Charles Howard Hinton. https://en.wikipedia.org/wiki/Charles_Howard_Hinton. Accessed 12 Oct. 2021.
16. Neve, A. (1926). Miss Annie Neve's reminiscences of Mrs. Clarke-Keer. *Pharmaceutical Journal and Pharmacist, 117*, 374–375.
17. Hudson, B. (11 April. 2019). Keer [née Clarke], Isabella Skinner Clarke (1842–1926). *Oxford Dictionary National Biography*. https://doi.org/10.1093/odnb/9780198614128.013.369105. Accessed 11 Jan. 2021.
18. Anon. (1905). Obituary: Lucy Everest Boole. *Proceedings of the Institute of Chemistry of Great Britain and Ireland, 29*(Part II), 26.
19. Shellard, E. J. (1982). Some early women researcher workers in British pharmacy 1886–1912. *Pharmaceutical Historian, 12*(2), 2–3.
20. Holloway, S. W. F. (1919). *Royal pharmaceutical society of Great Britain 1841–1991: A political and social history*. The Pharmaceutical Press.
21. Henry, T. A. (1950). Wyndham Rowland Dunstan. 1861–1949. *Obituary Notices of Fellows of the Royal Society, 7*(19), 63–81.
22. Rayner-Canham, M. F., & Rayner-Canham, G. W. (2017). *A chemical passion: The forgotten saga of chemistry at British independent girls' schools, 1820–1940*. Institute of Education Press.
23. Anon. (31 May 2021). Antimony Potassium Tartrate. https://en.wikipedia.org/wiki/Antimony_potassium_tartrate. Accessed 15 Jan. 2021.
24. Dunstan, W. R., & Boole, L. E. (1889). Chemical observations on tartar emetic. *Pharmaceutical Journal, 19*, 385–387.
25. (Jan. 1892). *London School of Medicine for Women, Annual Report*.
26. (1894). *London School of Medicine for Women, Annual Report*.
27. Anon. (27 Apr. 2021). Croton Oil. https://en.wikipedia.org/wiki/Croton_oil. Accessed 12 Jan. 2021.
28. Dunstan, W. R., & Boole, L. E. (1895). An enquiry into the nature of the vesicating constituent of croton oil. *Proceedings of the Royal Society of London., 59*, 237–249.
29. Hecker, E. (1967). Phorbol esters from croton oil: Chemical nature and biological activities. *Naturwissenschaften, 54*(11), 282–284.
30. (1898). *London School of Medicine for Women, Annual Report*, p. 17.
31. Rayner-Canham, M. F., & Rayner-Canham, G. W. (2003). Pounding on the doors: The fight for the acceptance of British women chemists. *Bulletin for the History of Chemistry, 28*(2), 110–119.
32. Anon. (16 Jun. 1894). Reports. *Proceedings of the Institute of Chemistry of Great Britain and Ireland*, 192.
33. Anon., (Oct. 1905). Obituary: Miss Boole. *Magazine of the London (Royal Free Hospital) School of Medicine for Women, 2*, 454–455.
34. Blackett, L. M. (1905). Memorial to Miss Boole. *Magazine of the London (Royal Free Hospital) School of Medicine for Women, 31*, 501.

# 12 Clare de Brereton Evans

*In 1897 as a result of Lucy Boole's ill-health (see* Chap. 11*), the Council of the London School of Medicine for Women had reduced her responsibilities. Clare de Brereton Evans was then appointed Lecturer in Chemistry to take over part of Boole's duties. Evans continued on the Chemistry staff of the School until 1912, when she decided to focus solely on research at University College, London.*

## Early Life of Clare De Brereton Evans

Clara (Clare) de Brereton Evans was born in 1866 in Bath to William de Brereton Evans, Surgeon and retired Deputy Inspector General of Hospitals in Madras (India), and Emma Soames. Youngest of seven children, Evans was first educated at the Royal School for Daughters of Officers, Bath. The School's mission was to provide practical and religious education for the daughters of army officers who might otherwise be unable to afford the fees of private schooling [1].

From that School, Evans entered Cheltenham Ladies' College (CLC) in 1885. The CLC had particularly good science teaching laboratories [2], such that students at the College were permitted to study there towards a B.Sc. (External) degree of the University of London [3]. Evans was awarded her external B.Sc. (London) at the CLC in 1889.

## Teaching at CLC

For at least the next two years, Evans remained at CLC as Teacher of Chemistry.

She must have been highly regarded among the Teachers of Chemistry in independent girls' Schools because Evans was invited to write a chapter, "The Teaching of Chemistry," in a book: *Work and Play in Girls' Schools by Three Head Mistresses*



M. Rayner-Canham and G. Rayner-Canham, *Pioneers of the London School of Medicine for Women (1874–1947)*, Perspectives on the History of Chemistry,
https://doi.org/10.1007/978-3-030-95439-0_12

[4]. This book contained a compilation of articles by educational experts in their fields.

In her chapter, Evans expresses her belief that junior, as well as senior, schoolgirls needed exposure to practical chemistry [5]:

> For success in examinations it is now necessary to have a certain amount of practical knowledge of chemistry, and examination classes are therefore given some practical training, but this reform still remains to be extended universally to the junior classes, which need even more than the senior ones that the teaching should be objective: a child may learn and repeat correctly a dozen times that water is composed of oxygen and hydrogen, and the thirteenth time she will assure you that its constituents are oxygen and nitrogen; but let her make the gases for herself, test them and get to know them as individuals, and mistakes of this kind will become impossible.

## Research with Henry Armstrong

Evans moved to London in 1891 to undertake research with Henry Armstrong at the Central Technical College, South Kensington, later part of Imperial College. Her research with Armstrong focussed upon chemical reactions of benzenoid amines, the first publication being on the derivatives of dimethylaniline [6], and the second on reactions of diethylaniline, dimethylortho-toluidine, and dimethylpara-toluidine [7].

Evans also synthesized enantiomorphs (mirror image crystals) of ethylpropylpiperidinium iodide [8]. This research was of special significance, as was commented in a history of the Chemistry Department of Imperial College [9]:

> Some impressive crystallographic work was carried out also by another of Armstrong's research students, Clare de Brereton Evans. She made mirror image crystals of ethylpropylpiperidinium iodide but found that neither form exhibited optical activity in solution. She continued to work on this problem, not resolved until it was recognised that asymmetry can be associated with pentavalent nitrogen.

As a result of her research, Evans was awarded a D.Sc. in 1897, the first woman to receive that degree from the University of London.

## Teaching at LMSW and Research with Ramsay

Evans was appointed Lecturer at the London School of Medicine for Women (LSMW) in 1897 [10] (Fig. 12.1). She combined her LSMW teaching with part-time research at nearby University College, London, under Sir William Ramsay. Ramsay had many of his research group searching for new chemical elements. He was convinced that the thorium-containing mineral, thorianite, a sample of which he had obtained from Ceylon (now Sri Lanka) contained one or more unknown elements in trace amounts [11].

**Fig. 12.1** Students in the Chemistry Laboratory, 1900. Public domain: London Metropolitan Archives, unknown photographer

In her separations, Evans obtained a small amount of an unidentified dark brown sulphide. To have enough product to identify the substance, she commenced with 80 kg of thorianite, which gave about 150 g of the unknown sulphide. This compound she purified, then reduced to the possible newly-discovered metallic element [12].

Evans attempted to determine the atomic weight of the element by means of electrolytic deposition. Working with Otto Brill, a Czech researcher with Ramsay, Evans devised a modification to the Nernst microbalance so that it could measure to as little as 0.0004 g, thus enabling only a tiny sample to be used [13]. However, the results from this analytical method, and from spectroscopic analysis, proved inconclusive.

## Professional Activities

Like Boole (see Chap. 11), Evans was one of the 19 signatories of women chemists who unsuccessfully petitioned for admission to the Chemical Society [14]. Evans was also one of 31 women chemists who, in 1909, submitted a letter to the *Chemical News* [15]. The letter was a response to a report claiming that the women petitioners were linked to radical elements in Society promoting political enfranchisement. In the letter, the women chemists made it clear that the sole link between them was a common interest in chemistry.

## Later Life

Evans continued teaching chemistry at LSMW until 1912. Her departure was reported in the *LSMW Magazine* [16]: "Very much regret is also felt at the resignation of Miss C. Evans from the Lectureship in Chemistry. It is understood that Miss Evans desires to devote herself more completely to the research work in which she has always been engaged."

At some later unknown date, Evans changed her interest from research chemistry to early Christian mystical writings. In particular, she translated the works from German to English of the thirteenth-century German theologian, philosopher, and mystic, Meister Eckhart (Eckhart von Hochheim) [17, 18]. Nothing else is known of Evans later life, except that she continued residing at the same address of 47 Campden Hill Court, London, until her death on 10 August 1935, aged 69.

## Commentary

It is unfortunate that very little documentation survives about de Brereton Evans. Fortunately, for her successor, Sibyl Widdows, we have a much fuller account. It is the life and work of Widdows which is the subject of Chap. 13.

## References

1. Osborne, H., & Manisty, P. (1966). *A history of the Royal School for daughters of officers of the Army, 1864–1965*. Hodder & Stoughton.
2. Rayner-Canham, M. F., & Rayner-Canham, G. W. (2017). *A chemical passion: The forgotten saga of chemistry at British independent girls' schools, 1820–1940*. Institute of Education Press.
3. Clarke, A. K. (1953). *A history of the Cheltenham Ladies' College, 1853–1953*. Faber.
4. Evans, C. de B. (1898). The teaching of chemistry. In D. Beale, L. H. M. Soulsby, & J. F. Dove (Eds.), *Work and play in girls' schools by three head mistresses* (pp. 307–319). Longmans, Green, and Co.
5. Ref. 4, Evans, pp. 310–311.
6. Evans, C. de B. (1895–96). Researches on tertiary benzenoid amines: I. Derivatives of dimethylaniline. *Proceedings of the Chemical Society, London, 11*, 235–236.
7. Evans, C. de B. (1896–97). Researches on tertiary benzenoid amines II. *Proceedings of the Chemical Society, London, 12*, 234–235.
8. Evans, C. de B. (1897). Studies on the chemistry of nitrogen. Enantiomorphous forms of ethylpropylpiperidonium iodide. *Journal of the Chemical Society, Transactions,* (Pt. 1), *71*, 522–526.
9. Gay, H., & Griffith, W. P. (2017). *The chemistry department at Imperial College London: A history, 1845–2000* (p. 95). World Scientific Publishing.
10. (1898). *London School of Medicine for Women, Annual Report*, p. 17.
11. Fontani, M., Costa, M., & Orna, M. V. (2015). *The lost elements: The periodic table's shadow side* (pp. 219–223). Oxford University Press.
12. Evans, C. de B. (1908). Traces of a new tin group element in thorianite. *Journal of the Chemical Society, Transactions, 93*, 666–668.

13. Brill, O., & Evans, C. de B. (1908). The use of the micro-balance for the determination of electrochemical equivalents and for the measurement of densities of solids. *Journal of the Chemical Society, Transactions, 93*, 1442–1446.
14. Rayner-Canham, M. F., & Rayner-Canham, G. W. (2003). Pounding on the doors: The fight for the acceptance of British women chemists. *Bulletin for the History of Chemistry, 28*(2), 110–119.
15. Beveridge, H.H., et al. (5 Feb. 1909). Women and the fellowship of the Chemical Society. *Chemical News*, 70.
16. Anon. (Mar. 1912). Hospital and school news. *Magazine of the London (Royal Free Hospital) School of Medicine for Women, 8*, 77.
17. Evans, C. de B. (translator). (1924). *Meister Eckhart by Franz Pfeiffer*. John M. Watkins.
18. Evans, C. de B. (translator). (1931). *The works of Meister Eckhart, Doctor Ecstaticus*. John M. Watkins. reprinted 1952.

# Sibyl Taite Widdows

# 13

*It was in 1901 that Sibyl Taite Widdows was appointed Demonstrator in Chemistry at the London School of Medicine for Women (LSMW), working under Clare de Brereton Evans. Widdows was the longest-term staff member of the Chemistry Department, and the one about whom the most information could be found.*

## Early Life

Sibyl Taite Widdows (Fig. 13.1) was born on 27 May 1876 in Lewisham, Kent. Her father was Thomas Widdows, Solicitor, and her mother, Elizabeth Shoosmith. She had one sister and two brothers. Widdows attended Dulwich High School for Girls. In 1896, Widdows entered Royal Holloway College (RHC), one of the women's constituent Colleges of London University [1], graduating with a B.Sc. in Chemistry in 1900.

## Teaching at LSMW

Widdows was to dominate the Chemistry Department at the London School of Medicine for Women (LSMW) for 40 years. In 1901, she had been appointed as Demonstrator in Chemistry. Then, as reported in the *LSMW Magazine*, she was promoted to Teacher of Practical Chemistry in 1904 following the death of Boole [2]:

> The appointment of Teacher of Practical Chemistry, left vacant by the death of Miss Boole, has been given to Miss Widdows, B.Sc., who has since 1901 assisted Miss Boole in the work of the Laboratory. We feel sure that the students will very heartily welcome this appointment.



M. Rayner-Canham and G. Rayner-Canham, *Pioneers of the London School of Medicine for Women (1874–1947)*, Perspectives on the History of Chemistry,
https://doi.org/10.1007/978-3-030-95439-0_13

**Fig. 13.1** Sibyl Widdows. Public domain: London Metropolitan Archives, unknown photographer

For the 1911–1912 Examinations, the Examiner for inorganic chemistry was Clare de Brereton Evans; for practical chemistry, Widdows; and for organic chemistry, Evans and Widdows, jointly [3]. Widdows was promoted to Lecturer in 1912, following the departure of Evans, then appointed Head in 1935 [4]. Widdows, like Boole and Evans, was a signatory to the unsuccessful petition by 19 women chemists for admission of women to the Chemical Society [5]. And, like Evans, Widdows was another of the 31 signatories to the 1909 letter to *Chemical News* [6].

Life in the Chemistry Department at LSMW was described in an Obituary of Widdows in the *LSMW Magazine* by her successor, Phyllis Sanderson [7]:

> Of miniature stature, alert and sprightly, Miss Widdows possessed such vitality and drive that it seemed a store of dynamite must be housed within her small frame. As with all who have a gift for it, she loved teaching and did so with untiring verve, never despairing even of the lowest of her flock. ... Practical classes, certainly no playtime, held an element of excitement (possibly mixed with terror) that kept everyone on their toes; for S.T.W. would systematically work her way down the laboratory, visiting student after student to ensure that each in turn was fully understanding what they were doing. Suddenly a loud scream of dismay would ring out and all would shudder, knowing full well that some unfortunate student had uttered an appalling chemical howler or had committed some dangerous crime such as heating an inflammable liquid with a naked flame. Near neighbours of the offender would immediately rush off to recharge their wash-bottles or busy themselves at the fume-cupboard hoping (in vain) to escape the deadly searching questions so soon to reach them. Just as frequently there were roars of laughter at the odd joke or cries of triumph as she found one of her flock had at last understood some basic chemical principle.

## Student Verse About Widdows

Students included Widdows in their poetic verse published in the *LSMW Magazine*. This first one shown is part of a lengthy set of verses covering many of the staff members of the time [8]:

> A general Chemmy favourite is Miss Widdows, B.Sc.,
> She mothers all the students and invites them all to tea,
> So why not all be medicals—and she might ask you too,
> And put you through your paces at the L.S.M.W.

This second set of verses pertains specifically to Widdows [9]:

> Come all you young chits, take leave of your wits,
> For to Chemistry you must hie,
> And every pair must a test prepare
> With the lecturer standing by.
> Take 20 grams. or more
> Of $H_2SO_4$
> And all the girls will drip it, drip it, drip it on the floor.
> "I'm curst," says Hurst, "my test tube's burst,
> 'Twas Miss Widdows that told me wrong!"
> "O Larks!" says Fox, "I'm on the rocks!"
> And so says everyone.
> Miss Widdows then begins
> To tell them of their sins,
> And the salty tears they drip it, drip it, drip it down their chins.

## Research Activities

In addition to teaching, Widdows was an active researcher. Her first publication, dated 1906, was a study of the rate of absorption of chloroform, the research being undertaken with Dr. T. G. Brodie, F.R.S., Lecturer on Physiology at the London School of Medicine for Women. In the introduction, they explained the purpose of the research [10]:

> The object of the following experiments was to determine the rate of absorption of chloroform during the induction of anaesthesia. All the experiments have been conducted upon cats, the plan adopted being to make the animal inspire air of known chloroform content and to collect the expired air and determine the amount of chloroform it contained.

Widdow's research then turned towards synthetic organic chemistry. Two years later, she was the co-author of a 12-page paper on substituted pyridines, the other author being William Mills (mentioned above) [11]. This investigation focussed upon the unique behaviour of the 2-substituted pyridine derivatives, as they explained [12]:

> The 3-substituted derivatives of pyridine show a far-reaching analogy to the corresponding compounds of benzene. This is, however, not the case with the 2-derivatives; in these the substituent radicle appears to become deprived, in many respects, of its aromatic character through the closer influence of the nitrogen.

Continuing her research in organic chemistry, Widdows next publication (1914) was with Ida Smedley Maclean. Maclean held a Beit Research Fellowship at the Lister Institute of Preventative Medicine, London [13]. Though Maclean's main research was on fat metabolism and synthesis, the research with Widdows was on unexpected products from a Grignard reaction [14]:

> In an investigation which had for its objective the preparation of an optically active compound, we desired as starting materials the *p*-carbethoxy- and *p*-dimethylamino-derivatives of ββ-diphenylpropiophenone. It was expected that these substances would be produced by the action of magnesium phenyl bromide respectively on the *p*-carbethoxy- and *p*-dimethylamino-derivatives of phenyl styryl ketone. These reactions have been investigated. As the behaviour of these substances in subsequent reactions did not lead to the results anticipated, it has been found necessary to modify our original scheme, and we desire therefore at this stage to record the properties of the substances prepared in the course of this investigation.

The following year, Widdows co-authored a research paper with stereochemist, Alex McKenzie. At the time of publication, McKenzie was Head of the Chemistry Department, University College, Dundee [15]. However, from 1902 until 1905, he had been Assistant Lecturer at Birkbeck College, London University, which was where the two probably met. This research studied the racemization of phenyl-p-tolylacetic acid in alcoholic potassium hydroxide [16].

At the beginning of the twentieth century, nearly all the pharmaceuticals sold in Britain were manufactured in Germany. With the start of the First World War, this supply was cut off. All University Chemistry Departments in Britain were contacted requesting them urgently to turn their efforts to the production of the needed chemicals [17]. Widdows was recruited to join the all-women team at Imperial College, London University, to synthesize pharmaceuticals under the direction of Martha Whiteley [18].

## Professional Activities

Women were finally admitted to the Chemical Society in 1920. Immediately following the change in rules, twenty-one women chemists were elected to Fellowship, Widdows being one of them [19]. Her prestigious nominators included Jocelyn Thorpe and James C. Philip of Imperial College, London; and William Mills, then of the University of Cambridge, with whom Widdows had undertaken research at the Northern Polytechnic Institute, London in 1908.

In 1922, Widdows, together with Winifred C. Cullis, Head of the Department of Physiology at LSMW, penned a letter to the magazine *Time and Tide* about the School. *Time and Tide* had been launched in 1920 as the only weekly review magazine owned and edited by a woman, Margaret Haig Thomas, 2nd Viscountess

Rhondda. Thomas sought out feminists, radical women writers, and journalists at a time when the first stage of women's suffrage had been won. In the letter, Cullis and Widdows reminded readers of the important role of the London (Royal Free Hospital) School of Medicine for Women [20]:

> This hospital [The Royal Free Hospital], it should never be forgotten, was the very first to open its doors to women; it opened them not when it was a popular thing to do during a great European war, but when every other teaching hospital closed its doors and kept them closed until nearly forty years later. This hospital at the present, as in the past, is doing more for the medical education of women than any other teaching hospital.

## Research Studies on Human Lactation

Perhaps her experience at the Imperial College Laboratories gave Widdows the courage to embark on independent research. During the 1920s, she founded her own research field: that of blood and milk composition during human lactation. Widdows research was described in the *LSMW Magazine* [21]:

> Miss Widdows (Chemistry Department) is determining the calcium content of the blood under various conditions, to see what may be the limit of physiological variation during menstruation and pregnancy. She is hoping to extend these determinations to various pathological conditions, with a view to finding whether the calcium content of blood may be of use as a diagnostic factor.

In 1922, Widdows had a Letter published in *The Lancet* which gave a preview of her findings. The Letter opened with the following [22]:

> Sir, During the last two years I have been working on the calcium content of the blood of women during the different stages of pregnancy. Interesting variations have been observed here, which led me also to investigate the calcium content of the blood in conjunction with menstruation. In view of your recent leading article in *The Lancet* (Nov. 4$^{th}$) on this subject my figures may be of interest to you.

Widdows research on lactation must have become known around the LSMW, as the following satirical poem illustrates [23]:

**Sandwiches** [It has been reported that while on holiday recently a chemistry lecturer fed a cow on jam sandwiches.]

> I met a chemmy lecturer,
> I said: "Pray tell me now
> Your reasons for administ'ring
> Jam sandwich to a cow.
> This dietetic enterprise
> We really must insist
> Encroaches on the regions of
> The physiologist."
> "Was it because you thought that jam

In sandwiches would show
A stimulating action on
A young lactating cow?
Or else, perchance, 'twas but a bribe,
And while she chewed the cud
You hastened to your chemmy lab
With samples of her blood;
And thus secure of bovine love
You hoped to work out how
Her calcium would vary if
She had a baby cow?"
She said: "Your reasons are not right,
Not one of all the bunch!
The explanation's simply this—
*I couldn't eat my lunch.*"

The lactation research was published in two parts, the first in 1923 relating to normal pregnancies [24]; and the second in 1924 in cases where previous pregnancies had exhibited abnormalities [25]. In addition, Widdows turned her attention to breast milk and other mammary secretions, as this Letter to the Editor of the *British Medical Journal* described in 1931 [26]:

> For some time at this school a group of workers has been investigating breast milk, from both biochemical and chemical aspects. ... It has now been decided that this investigation should be extended to include secretions occurring before parturition, during menstruation, and other instances of mammary activity. As such cases are infrequent, may we ask the help of your readers in giving us the opportunity of getting into touch with women in whom the breasts become active before parturition, or independently of pregnancy?

All these later publications on women's mammary secretions had Margaret Frances Lowenfeld as co-author. Lowenfeld was more renown as a child psychologist and psychotherapist [27]. They obtained their samples from women at the Royal Free Hospital, Shoreditch Carnegie Welfare Settlement, and from the Mothercraft Training Society, Cromwell House, Highgate. Widdows and Lowenfeld, sometimes together with other co-researchers, produced an amazingly high productivity of publications over the 1927–1935 period [28–33].

## Later Years

Widdows retired in 1942 and died in London on 4 January 1960. In the lengthy obituary, Sanderson noted how Widdows had been [7]: "one of the last of the remarkable women who staffed the School during the first forty years of this century, an uphill and critical period in the history of this medical School." Phyllis Sanderson added [7]:

> As so many of her contemporaries, she was an ardent feminist and willingly sacrificed her own career as a chemist for the cause most dear to her heart, the training of women doctors

> at Hunter Street, the only training ground in Medicine open to women in England at that time.

An anonymous Obituary in the *Journal of the Royal Institute of Chemistry* noted [4]:

> Miss Widdows was a magnificent teacher and devoted her life at the London School of Medicine to teaching and to analytical work in connection with research. She was a pioneer among woman at the school, and was much loved by students and colleagues.

## Commentary

Throughout this Chapter, we have shown that Widdows had a dominating presence in the Chemistry Department for several decades. In Chap. 14, we look at the life and work of Phyllis Sanderson and Anne Ratcliffe, the last two woman chemists to head the Chemistry Department before it was absorbed into that of the Royal Free Hospital.

## References


1. Rayner-Canham, M. F., & Rayner-Canham, G. W. (2017). *A chemical passion: The forgotten saga of chemistry at British Independent girls' schools, 1820–1940*. Institute of Education Press.
2. Anon. (1904). School notes. *Magazine of the London (Royal Free Hospital) School of Medicine for Women, 1*, 460.
3. Anon. (Mar. 1912). Hospital and school news: School examinations, *Magazine of the London (Royal Free Hospital) School of Medicine for Women, 8*, 80–81.
4. Anon. (1960). Sibyl Taite Widdows. *Journal of the Royal Institute of Chemistry, 84*, 233.
5. Rayner-Canham, M. F., & Rayner-Canham, G. W. (2003). Pounding on the doors: The fight for the acceptance of British women chemists. *Bulletin for the History of Science, 28*(2), 110–119.
6. Beveridge, H. H., et al. (5 Feb. 1909). Women and the fellowship of the chemical society. *Chemical News, 70*.
7. Sanderson, P. M. (1960). Obituary: Sibyl Widdows. *Royal Free Hospital Journal, 23*, 21–22.
8. Anon. (Mar. 1923). Hullo, everybody. *Magazine of the London (Royal Free Hospital) School of Medicine for Women, 18*(84), 37–38. [section on p. 38].
9. Anon. (Mar. 1923). Come all you young chits. *Magazine of the London (Royal Free Hospital) School of Medicine for Women, 18*(84), 38–39.
10. Brodie, T. G., & Widdows, S. T. (14 Jul. 1906). Preliminary report on the rate of absorption of chloroform during the induction of anaesthesia. *British Medical Journal, 2*, 79–83.
11. Mann, F. G. (1960). William Hobson Mills. 1873–1959. *Biographical Memoirs of Fellows of the Royal Society, 6*, 200–225.
12. Mills, W. H., & Widdows, S. T. (1908). Benzeneazo-2-pyridone. *Journal of the Chemical Society, Transactions, 93*, 1372–1384.
13. Rayner-Canham, M., & Rayner-Canham, G. (2011). Forgotten pioneers. *Chemistry World, 8*(12), 41.
14. Maclean, I. S., & Widdows, S. T. (1914). The action of magnesium phenyl bromide on derivatives of phenyl styryl ketone. *Journal of the Chemical Society, Transactions, 105*, 2169–2175.
15. Roger, R., & Read, J. (1952). Alexander McKenzie. 1869–1951. *Obituary Notices of Fellows of the Royal Society, 8*(21), 206–228.

16. McKenzie, A., & Widdows, S. T. (1915). Racemization of phenyl-*p*-tolylacetic acid. *Journal of the Chemical Society, Transactions, 107*, 702–715.
17. Rayner-Canham, M. F., & Rayner-Canham, G. W. (1999). British women chemists and the First World War. *Bulletin for the History of Chemistry, 23*, 20–27.
18. Whiteley. M.A. to Conway, A.E. (3 Oct. 1919). Letter. In *Women's war work collection.* Imperial War Museum.
19. Anon. (1920). Certificates of candidates for election at the ballot to be held at the Ordinary Scientific Meeting on Thursday, December 2nd. *Proceedings of the Chemical Society* (pp. 82–100).
20. Cullis, W. C., & Widdows, S. T. (17 Mar. 1922). Correspondence: Deeds not words. In *Time and tide.* Archives of the Royal Free Hospital, S. T. Widdows file.
21. Anon. (Mar. 1921). Research work at the LSMW. *Magazine of the London (Royal Free Hospital) School of Medicine for Women, 16*(78), 78.
22. Widdows, S. T. (11 Nov. 1922). The menstrual cycle and calcium content of the blood (Letter). *The Lancet, 200*, 1037.
23. Anon. (Jul. 1923). Sandwiches. *Magazine of the London (Royal Free Hospital) School of Medicine for Women, 19*(85), 117–118.
24. Widdows, S. T. (1923). Calcium content of blood during pregnancy. *Biochemical Journal, 17*, 34–40.
25. Widdows, S. T. (1924). Calcium content of blood during pregnancy II. *Biochemical Journal, 18*, 555–561.
26. Widdows, S. T. (16 May 1931). Aberrant mammary secretion. *British Medical Journal, 1*, 16.
27. Urwin, C. (23 Sept. 2004). Lowenfeld, Margaret Frances Jane (1890–1973). *Oxford Dictionary of National Biography*. https://doi.org/10.1093/ref:odnb/60993. Accessed 5 Feb. 2021.
28. Lowenfeld, M. F., Widdows, S. T., Bond, M., & Taylor, E. I. (1927). A study of the variations in the chemical composition of normal human colostrium and early milk. *Biochemical Journal, 21*, 1–15.
29. Lowenfeld, M. F., & Widdows, S. T. (1928). Researches on lactation. *Journal of Obstetrics and Gynaecology, 35*, 114–130.
30. Widdows, S. T., Lowenfeld, M. F., Bond, M., & Taylor, E. I. (1930). A study of the composition of human milk in the later periods of lactation and a comparison of that with early milk. *Biochemical Journal, 24*, 327–342.
31. Widdows, S. T., & Lowenfeld, M. F. (1933). A study of the composition of human milk. The influence of the method of extraction on the fat percentage. *Biochemical Journal, 27*, 1400–1410.
32. Widdows, S. T., Lowenfeld, M. F., & Chodak, H. H. (3 Nov. 1934). Percentage of fat in human milk. Influence of the method of extraction. *The Lancet, 224*, 1003–1004.
33. Widdows, S. T., Lowenfeld, M. F., Bond, M., Shiskin, C., & Taylor, E. I. (1935). A study of the antinatal secretion of the human mammary gland and a comparison between this and the secretion obtained directly after birth. *Biochemical Journal, 29*, 1145–1166.

# Phyllis Sanderson and Anne Ratcliffe 14

*In this chapter, we cover the lives of the last two Senior Staff of the Chemistry Department: Sanderson and Ratcliffe. Phyllis Sanderson was first hired as a Demonstrator in Chemistry in 1925. Then in 1942, she was appointed as Sibyl Widdow's successor as Head of Chemistry, following Widdow's retirement. Sanderson's successor, Anne Radcliffe was hired in 1929, her first appointment being as Demonstrator in Chemistry, and then she followed Sanderson through the appointment ranks.*

## Phyllis Sanderson

Phyllis Mary Sanderson (Fig. 14.1) was born in 1901 at Hove, Sussex, the middle of five children of Robert Sanderson, surgeon, and Agnes Mary Cooke. She was educated at Brighton and Hove High School. Sanderson completed her B.Sc.(Hons.) in Chemistry at University College, London, in 1924 [1].

Sanderson then spent a year of postgraduate study at the Westminster Children's Hospital, London. It was here that she undertook her first research, this being on infantile diseases. The investigation was undertaken jointly with Lucy Wills, Chemical Pathologist at the Royal Free Hospital, and Donald Paterson, Physician to Outpatients, at the Hospital for Sick Children, Great Ormond Street. It was hypothesized that, as calcium compounds are more soluble under acid conditions, gastric acidity might lead to better calcium ion absorption and hence a reduction in the incidence of rickets. However, the contrary was true, infants fed 'hydrochloric acid milk' absorbed less calcium. The research was published in 1926 [2].

Sanderson was appointed Demonstrator in Chemistry at the London School of Medicine for Women (LSMW) in 1925 [3], following the resignation of Marjory Wilson Smith (see Chap. 15). That same year, she was elected Fellow of the



M. Rayner-Canham and G. Rayner-Canham, *Pioneers of the London School of Medicine for Women (1874–1947)*, Perspectives on the History of Chemistry,
https://doi.org/10.1007/978-3-030-95439-0_14

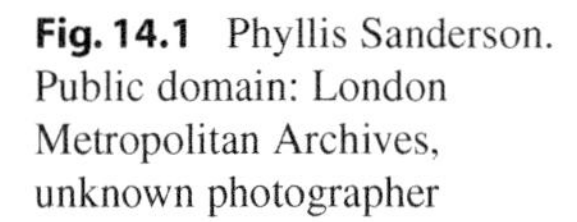

**Fig. 14.1** Phyllis Sanderson. Public domain: London Metropolitan Archives, unknown photographer

Chemical Society, her three nominators being chemistry faculty at University College, London: Frederick Donnan, Professor of General Chemistry; William Garner, Assistant Lecturer in Chemistry; and Henry Terrey, Reader in Chemistry.

## Sanderson's Later Academic Life

Sanderson was promoted to Senior Demonstrator in 1933 and to Assistant Lecturer in 1934. While continuing to teach at the LSMW, during the 1930s, she undertook research with Vincent Briscoe [4] at Imperial College, London. The study of industrial dusts, especially chemical aspects of silicosis in miners, resulted in three co-authored publications in a specialist mining journal between 1936 and 1937 [5–7], together with a summary in the journal *Nature* [8]:

> With the assistance and encouragement of the Institution of Mining and Metallurgy and the Medical Research Council, we have continued an investigation, initiated four years ago under the auspices of the Institute of Chemistry, upon the chemical nature of dusts causing silicosis. Using the salicylic acid dust filters described in a recent communication, we have now obtained, probably for the first time, analysable samples of mine dust as breathed by the miner. The dust was produced by wet drilling and blasting of granite in a Cornish mine.

Briscoe was awarded a medal by the Institute of Mining and Metallurgy for the method of dust analysis [9] while Sanderson received a Diploma of Imperial College for this work and then in 1939, a Ph.D. from University College, London for her research on solubility of silica and silicates.

Nothing could be found on Sanderson's activities during the Second World War. It is probable that she rejoined Briscoe's group. Briscoe worked for MI5 on the development of chemically sophisticated invisible ink formulations. In an account of the Imperial College Chemistry Department activities during the war, it was noted [10]: "Briscoe engaged also in a range of other covert activities that are difficult to pin down." If Sanderson was associated with his wartime group, it is likely details will never be known.

Sanderson was promoted to Lecturer at LSMW only in 1946. In a quiz 'Spot the Staff', in a 1954 issue of the *RFH Magazine*, one quote reads: "I start every day with a vitamin capsule and a du Maurier [a brand of cigarettes]." The key on a subsequent page identified the speaker as Dr. Sanderson [11].

Sanderson's first post-War publication was a historical review of the importance of urea in the history of organic chemistry. It was co-authored with Frederick Kurzer, a Lecturer in Chemistry at the Royal Free Hospital School of Medicine who had been hired in 1949. Kurzer, too, had an interest in the history of chemistry. The review was published in 1956 [12] and Sanderson and Kurzer followed the historical review with two research papers on the synthesis of some urea derivatives [13, 14].

Kurzer was one of the two co-authors of Sanderson's Obituary in the *Royal Free Hospital Journal* (*RFH Journal*). He referred in detail to Sanderson's enthusiasm for the history of science [1]:

> Dr Sanderson loved to delve into the history of chemistry and scientific thought in general. It was typical of her sense of justice that in one of these studies she should have rescued from oblivion a hitherto obscure 18th Century scientist William Cruickshank, by re-establishing his claims to several important discoveries that had been erroneously ascribed to another investigator. Not the least of the results of these efforts was her familiarity with all the great libraries of London. Her lively tales of the peculiarities of their arrangement and procedure, and the idiosyncrasies of both librarians and readers were a source of much amusement to her friends.

The research on Cruickshank, also co-authored by Kurzer, was published in 1957 [15].

Sanderson and Kurzer then changed their research direction to that of organic heterocyclic chemistry. Again, this was noted in her Obituary in the *RFH Journal* [1]: "Her patient and painstaking work established the necessary background of knowledge which helped in solving the chemical structure of certain types of heterocyclic compounds, the nature of which had been doubtful since their discovery in 1890." Giving her affiliation now as Imperial College, from 1957 to 1963, Sanderson and Kurze co-authored seven publications in heterocyclic synthesis [16–20].

Sanderson was a dedicated and inspiring teacher. Her Obituarists in the *RFH Journal* commented [1]:

> Her interest in the progress and welfare of the students was immense and was such that she became closely acquainted with each individual in a very short space of time. ... In her teaching, Dr. Sanderson's humour and patience contributed greatly to making a less popular

> subject more palatable and her classes were a happy combination of informality and discipline. Her lectures were a model of clarity, prepared with great care and always alive to new developments without neglecting established principles. She insisted on an understanding of the subject rather than the accumulation of facts. There is no doubt that her personality impressed itself in the lighter side of the life of the School: she was the perennial subject to be represented in the students' topical shows.

There is no information on Sanderson's last two years. She died on 7 September 1965 at the Royal Free Hospital.

## Anne Ratcliffe

Following Sanderson career-wise was Anne Ratcliffe, one of the last women Lecturers at the LSMW [21]. Born in London in April 1896, she obtained her qualifications at University College, London. Ratcliffe completed a B.Sc.(Hons.) in Chemistry in 1924 (as Annie Ratcliffe). Her initial appointment at LSMW in 1929, following the departure of Norah Laycock (see Chap. 15), was followed by promotion to Senior Demonstrator in 1940; Assistant Lecturer in 1945; Lecturer in 1947; and finally, Senior Lecturer in 1949.

Ratcliffe's research work focussed upon sterols found in fungi. This was the research field of John Addyman Gardner (see Chap. 15). In her sole publication, giving the St George's Hospital Medical School as her location, Ratcliffe discussed the reason for her choice of research [22]:

> It has been known for some time that many members of the fungus group contain mixtures of sterols and Mr J. A. Gardner thought that it might prove interesting to examine some of the commonly occurring fungi more thoroughly. The first fungus studied was *Boletus edulis*, which is utilized as food in Russia.

The results of the research showed that the sterols of *Boletus edulis* consist mainly of ergosterol together with small quantities of a sterol which closely resembles spinasterol. In 1939, Ratcliffe was awarded an M.Sc. based on her research into sterols and carbohydrates in certain fungi.

Upon Ratcliffe's retirement in 1961, Phyllis Sanderson wrote of her character in the *RFH Journal* [21]:

> That she is an inspired and tireless teacher was quickly realised by students ... Patient and kind though she is, however, Miss Ratcliffe would not tolerate shoddy work or bad manners ... She is one of those rare beings possessed of extreme intellectual honesty. Rather than risk passing on often erroneous textbook information to a student she would take infinite trouble reading original papers on the subject, and never would she say she understood anything unless she had probed to the depths and considered it from every possible angle.

No additional information could be found on Ratcliffe's life. She died in London, in July 1969, age 73.

## Commentary

Boole, de Brereton Evans, Widdows, Sanderson, and Ratcliffe, were the five senior figures of the LSMW Chemistry Department. However, there were others who filled the subsidiary, but no less important, roles in the Department. The final Chapter of this sequence will introduce the reader to the other women who contributed to the successful running of this women-chemist-only enclave.

## References


1. "M.H., & F.K". (1966). Obituary: Dr. Phyllis M. Sanderson. *Royal Free Hospital Journal, 27*, 190–191.
2. Wills, L., Sanderson, P., & Paterson, D. (1926). Calcium absorption in relation to gastric acidity (A study of rickets). *Archives of Disease in Childhood, 1*, 245–254.
3. Anon. (Nov. 1925). News. *Magazine of the London (Royal Free Hospital) School of Medicine for Women, 20*(92), 167.
4. Anon. (1961). Obituary: Henry Vincent Aird Briscoe. *Journal of the Royal Institute of Chemistry, 85*, 425.
5. Matthews, J. W., Holte, P. F., Sanderson, P. M., & Briscoe, H. V. A. (1936). Porous solid filters for sampling industrial dusts. *Journal of Chemistry Metallurgy Mining Society. South Africa, 37*, 161–166.
6. Briscoe, H. V. A., Matthews, J. W., Holte, P. F., & Sanderson, P. M. (1937). The sampling of industrial dust by means of the labyrinth. *Journal of Chemistry Metallurgy Mining Society. South Africa, 38*, 81–97, 227–229.
7. Briscoe, H. V. A., Matthews, J. W., Holte, P. F., & Sanderson, P. M. (1937). A note on some new characteristic properties of certain industrial dusts. *Journal of Chemistry Metallurgy Mining Society, South Africa, 38*, 145–147.
8. Briscoe, H. V. A., Matthews, J. W., Holte, P. F., & Sanderson, P. M. (1937). A note on some new characteristic properties of certain industrial dusts. *Nature, 139*, 753–754.
9. Gay, H., & Griffith, W. P. (2017). *The chemistry department of imperial college London: A history 1845–2000* (pp. 207–208). World Scientific Publishing.
10. Ref. 9, Gay & Griffith, p. 216.
11. Anon. (16 Mar. 1954). Spot the staff. *Royal Free Hospital Magazine, 15*, 23.
12. Kurzer, F., & Sanderson, P. M. (1956). Urea in the history of organic chemistry. Isolation from natural sources. *Journal of Chemical Education, 33*, 452–459.
13. Kurzer, F., & Sanderson, P. M. (1957). Urea and related compounds. Part IV. Some aromatic and aliphatic dithioformamidines. *Journal of the Chemical Society*, 4461–4469.
14. Kurzer, F., & Sanderson, P. M. (1959). Urea and related compounds. Part VI. sym-diaryldithioformamidines. *Journal of the Chemical Society*, 1058–1064.
15. Kurzer, F., & Sanderson, P. M. (1957). The work of William Cruickshank. *Chemistry and Industry*, 456–460.
16. Kurzer, F., & Sanderson, P. M. (1960). Thiadiazoles. Part X. The synthesis and isomerisation of 2-aryl-5-arylamino-3-arylimino-$\Delta^4$–1,2,4-thiadiazolines. *Journal of the Chemical Society*, 3240–3249.
17. Kurzer, F., & Sanderson, P. M. (1962). Heterocyclic compounds from urea derivatives. Part III. Synthesis and cyclisation of isothioureas derived from o-aminothiophenol and diarylcarbodiimides. *Journal of the Chemical Society*, 230–236.
18. Kurzer, F., & Sanderson, P. M. (1963). Heterocyclic compounds from urea derivatives. Part V. Synthesis and cyclisation of N-o-hydroxyphenyl-N′N″-diarylguanidines. *Journal of the Chemical Society*, 240–245.

19. Kurzer, F., & Sanderson, P. M. (1963). Thiadiazoles. Part XIII. Isomerisation of "Hector's bases. *Journal of the Chemical Society*, 3333–3336.
20. Kurzer, F., & Sanderson, P. M. (1963). Thiadiazoles. Part XIV. A link between sym-diaryldithioformamidines and "Hector's bases. *Journal of the Chemical Society*, 3336–3342.
21. "P. M. S". (1961). Anne Ratcliffe. *Royal Free Hospital Journal, 24*, 15.
22. Ratcliffe, A. (1937). The sterols and carbohydrates in fungi I *Boletus edulis*. *Biochemical Journal, 31*(2), 240–243.

# Other Chemistry Staff 15

*The past four chapters have provided individual accounts of the most prominent women chemists of the London School of Medicine for Women. They were supported by a series of women chemists holding junior positions in the department. Because the Chemistry Department was, in many ways, an add-on to the School, it was poorly documented, except for the Lecturers, thus the accounts are far less complete than we would have liked.*

## John Addyman Gardner

Over the years, there was one significant male member associated with the Chemistry Faculty at the London School of Medicine for Women (LSMW): John Addyman Gardner. In 1905, Gardner had been placed in charge of the chemical section of the Physiological Laboratory, Imperial Institute, South Kensington, though his long-term position was as Chemist at St. George's Hospital [1, 2]. In 1912, he was additionally appointed as Lecturer in Organic and Applied Chemistry at LSMW [3]: "We welcome several new members of the School staff this term. … Mr. J. A. Gardner, M.A., F.I.C., Lecturer in in Organic and Applied Chemistry …".

Nevertheless, Gardner's focus continued to be research at the other institutions. Some of the research was undertaken by May Williams (see below) and, later, some with Anne Ratcliffe (see Chap. 14). With the closure of the Physiological Laboratory after the First World War, Gardner continued his research at the Biochemical Laboratories of the St. George's Hospital Medical School. It is not clear when he ceased to be Lecturer at LSMW. In another obituary of Gardner, it simply states that he taught at LSMW [4]: "… for some years."



M. Rayner-Canham and G. Rayner-Canham, *Pioneers of the London School of Medicine for Women (1874–1947)*, Perspectives on the History of Chemistry,
https://doi.org/10.1007/978-3-030-95439-0_15

**Fig. 15.1** Chemistry Staff of the LSMW, 1916: from left to right, Mrs. Stirling-Taylor; Mrs. Matthews; Miss Widdows; Miss MacKenzie; and Miss M. Williams. Public domain: London Metropolitan Archives, unknown photographer

## Elsie Forrest

In addition to the women chemistry Lecturers discussed in Chaps. 11–14, there were several others appointed over the years as Demonstrators of Chemistry (Fig. 15.1). One of the earliest of these was Elsie Forrest. Forrest followed Lucy Boole as Demonstrator, when Boole was promoted in 1893 [5]: "Miss Forrest B.Sc. (London) was recommended to succeed Miss Boole as Demonstrator of Chemistry."

Forrest was born on 11 August 1871 in Macduff, Banff, Scotland. However, for her early life, the family lived in Edinburgh West Coates, Midlothian. Forrest's father was Robert Gibb, Minister of West Coates Parish Church, and her mother, Margaret Stephen. She graduated with a B.Sc. in Chemistry from the University of London in 1892. At age 29, the 1901 Census reported Forrest as being a teacher of Chemistry (presumably at LSMW). It must have been a scientific family, as her elder sister was reported as a Mathematics Tutor, and her younger sister also as a Teacher of Chemistry. In 1911, age 39, Forrest was listed in the Census as a Lecturer in Mathematics, with a South Kensington address. She died April 1946, age 74, in Basford, Nottinghamshire.

## Norah Ellen Laycock

In 1904, Norah Ellen Laycock was appointed Demonstrator of Chemistry as replacement for Forrest [6]. Laycock was born on 21 January 1877 at Keithley, Yorkshire, to Arthur Laycock, initially sawmill manager and later estate agent, and Ellen Elizabeth Scott. She was educated at James Allen School, Dulwich, then at Streatham Hill High School, London. Laycock entered Royal Holloway College (RHC) in 1897 and obtained her B.Sc. degree in 1901.

Her occupation, if any, is unknown, until her appointment to the Chemistry Department at the LSMW in 1904. Then in 1916, her position became a joint one shared between the Biology and Chemistry Departments and the following year changed again to be solely Demonstrator in Biology as noted in the *LSMW Magazine* [6]: "Miss N. E. Laycock B.Sc, Demonstrator in Chemistry has been appointed Demonstrator in Biology." Laycock was promoted to Assistant Lecturer in Biology in 1918 [7].

In 1929, Laycock was granted a leave of absence for two terms. This leave was for the purpose of working with the renowned educationalist, Isabel Fry, on the design and launch of an experimental School. Laycock must have quickly decided that her future opportunity was at Fry's experimental School, as later the same year, it was noted in the *LSMW Annual Report* that [8]: "After 25 years work in the School Miss Laycock B.Sc. assistant lecturer in Biology has resigned to take up the appointment of Headmistress of Mayortorne School, Bucks. The Council offer her their sincere thanks for her devoted service to the interests of the School."

It was actually named the Farmhouse School. In 1917, Isabel Fry, Quaker, social reformer and educationalist, had set up the Farmhouse School, occupying Mayortorne Manor in Buckinghamshire [9]. Her progressive views on education had been influenced by the work of John Dewey and Maria Montessori. The curriculum included economics, morality, and grammar. The School also ran a small dairy farming business, and students mixed farm work with conventional education. Laycock took over as Headmistress in 1929 and later purchased the School. She died there in November 1951, age 74.

## Yvonne M. D. Cooper

The Demonstrator in Chemistry who succeeded Laycock was Yvonne Margaret Dyson Cooper [7]. Cooper was born 2 October 1895, her father being Arthur Dyson Cooper, retired solicitor, and her mother, Mary Henrietta Cecilia Fulcher. In Cooper's early years, the family lived in Sutton, Surrey, but subsequently, they moved to France. She received a *certificat d'études* in 1907. This was a diploma awarded at the end of elementary education certifying that the student had acquired basic skills in writing, reading, mathematics, history, geography, and applied sciences. Cooper received her 1st and 2nd *baccelauréat* in 1910 and 1911, the French national academic qualification at the completion of secondary education.

Upon the family returning to England, Cooper spent two years at Portsmouth Municipal College then entered RHC in 1914. During her time at the RHC, she was quite athletic, her student register page noting she was involved in rowing, hockey, tennis, and was good at swimming. Cooper was also Secretary of the RHC Newspaper Club. In terms of personal comments, there was a margin note that [10]: "Rather out of things. Very little sense of humour. Seems to have no friends in her own year." Cooper graduated with a B.Sc. in 1917.

Cooper was appointed Demonstrator in Chemistry at LSMW in that same year of 1917. In 1939, Cooper was a French language teacher in Bournemouth and a supporter of the Association of Friends of the French Volunteers (AFV). It is unclear when she had left employment at the LSMW. Then in 1951, she co-authored a book, with Helena Vaughen Davies, titled: *Aids to Practical Hygiene for Nurses* [11]. Cooper died in Bournemouth on 14 February 1972.

## May Williams

May Williams was born on 7 May 1886, daughter of Ralph Williams, Clerk in Holy Orders and Vicar of St. Luke's Kilburn, Paddington, and Lucy Anne Tayler [12]. Educated at Notting Hill High School, Williams entered RHC in 1905. Among her many activities at RHC were Captain of the Boat Club, and President of the RHC Chemical Society. Williams seems to have been an extremely dedicated student. According to her student record, in January 1907, there was the report [12]: "Report from doctor. 'Less work, more fresh air & exercise & plenty of sleep.' If she continues to work hard she will have bad break-down. Must have another year at College, or give up working for Honours. Arranged that she should have a fourth year & work for Honours."

Williams completed a B.Sc. (Hons.) Chemistry in 1909. She was appointed as Demonstrator in Chemistry at LSMW in the same year, and promoted to Senior Demonstrator in 1920, and to Assistant Lecturer in 1921.

In 1921, she received an M.Sc. in Chemistry jointly from RHC and LSMW based on her research on quinoline derivatives with John Addyman Gardner. The same year, Williams was co-author of a publication with Gardner on the measurement of cholesterol in solutions [13]. In 1921, a detailed account of the research was given in the *LSMW Magazine* [14]:

> Miss Williams has been working with Mr. Gardner on two subjects. The Lieberman colour reaction of cholesterol and its isomers has been investigated qualitatively and quantitatively, with regard especially to the use of this reaction as a method of estimation of sterol in tissues. Some of the reactions of chloropicrin, a substance used as a poison gas during the War, have also been studied. Its use has been proved as a substitute for nitrobenzene in Skraup's synthesis of quinoline, and various homologues and derivatives have been prepared. The action of some reducing agents on chloropicrin is now being determined and also the use to which it may be put in the synthesis of certain dyes.

In fact, the synthesis of dyes must have been fruitful as a patent on quinoline-derived dyes was issued in 1922 in their joint names [15]. Based on her research, also in 1921, she was elected Fellow of the Chemical Society, her nominators being her colleague, Sibyl Widdows; her research supervisor, John Addyman Gardner; and two of her former professors at RHC, Moore and Philip [16].

At some date, Williams must have resigned from the LSMW staff in order to pursue full-time research at Imperial College. Williams was re-hired at the LSMW with the rank of Demonstrator in 1935, as it was noted in the *LSMW*

*Magazine* [17]: "We welcome Miss May Williams MSc AIC [Associateship of Imperial College] who has returned to the staff this session as Demonstrator in Chemistry."

At her retirement in 1946, it was reported in the *LSMW Magazine* [18]:

> Miss Williams brilliant gifts as a teacher, her renowned patience with the students to whom chemistry was no easy subject, and the great interest she took in all that concerned the welfare of the school, will be greatly missed. It is difficult to imagine the Senior Common Room and the Chemistry Department without her.

## Effie Isabel Cooke (Mrs. Stirling-Taylor)

Effie Isabel Cooke was born 20 May 1874 in Wood Green, Middlesex. Her parents were William Henry Cooke, barrister-at-law and solicitor, and Emily Amelia Davison. In 1903, Cooke married George Robert Stirling Taylor, barrister-at-law.

Stirling-Taylor was hired at LSMW in 1915, though it is not clear as to her position. She retired in 1936, the *LSMW Magazine* commenting [19]: "Mrs. Stirling Taylor is leaving us after 21 years of service to the School in the Department of Chemistry, and in the realms of Art and Literature." A brief mention of Stirling-Taylor's death was reported in a later issue of the *LSMW Magazine* [20]: "April 20th 1947 at Quarry Orchard, The Clears, Reigate, Effie Isobel, widow of G. R. Stirling-Taylor of Pump Court, London, E.C., aged 73."

## Marjory Wilson-Smith (Mrs. Farmer)

While Stirling-Taylor only commenced work after marriage, the career of Marjory Jennet Wilson-Smith ended with her marriage. Wilson-Smith was born on 19 May 1899 at Bath, daughter of Thomas William Wilson-Smith, physician, and Alice Maud Wilks. She was educated at Bath High School and Cheltenham Ladies' College before entering RCH in 1918 [20]. The Student Record noted her as [21]: "Intense, v. enthusiastic & conscientious." Wilson-Smith graduated with a B.Sc. (Hons.) in Chemistry in 1921.

From 1921 until 1925, Wilson-Smith was a Demonstrator in Chemistry at LSMW, though she spent the 1922–1923 year as a research student in the Organic Chemistry Department of Imperial College, London. In 1925, Wilson-Smith accepted an appointment with RHC as Assistant Lecturer and Demonstrator in the Chemistry Department. Two years later, she was elected Fellow of the Chemical Society, her nominators being faculty at Imperial College: Martha Whiteley, Jocelyn Thorpe, Ernest Farmer, and Kon [22].

Wilson-Smith resigned in 1930, as the *RHC College Letter* reported [23]:

> The last loss to be recorded is of Miss M. J. Wilson-Smith (now Mrs Farmer) for two years Assistant Lecturer and Demonstrator in the Chemistry Department. Shortly after the end of

> term we heard of her engagement to a fellow-scientist at Imperial College. Her marriage followed within a few weeks. Those who remember her record for industry will note, perhaps only with faint surprise, that, according to Rumour, Mrs Farmer has for the last two months done "not a stroke" of scientific work.

According to the obituarist of Ernest Farmer, Wilson-Smith worked with Farmer after marriage [24]: "During this period, in 1930, he married Marjorie Wilson-Smith, she was one of his research students and continued for many years to assist him with his researches." However, none of Farmer's later publications list Wilson-Smith as a co-author. She died in April 1977 in North London.

## Commentary

This chapter has carried the story of the LSMW—at least through the account of the women-run Chemistry Department—to the date of the demise of the School. Why did it end, and did any of the ethos carry over after its absorption into the Royal Free Hospital? These are the topics for the final chapter of this compilation.

## References


1. Ellis, G. W. (1947). Obituary notice. *Biochemical Journal, 41*, 321–324.
2. Ellis, G. W. (1947). Obituary notice. John Addyman Gardner. 1867–1946. *Journal of the Chemical Society*, 865–866.
3. Anon. (Nov. 1912). School notes. *Magazine of the London (Royal Free Hospital) School of Medicine for Women, 8*(53), 128.
4. Ellis, G. W. (1946). Obituary: Mr. J. A. Gardner, *Nature, 158*, 17.
5. Anon. (1894). *London School of Medicine for Women, Annual Report.*
6. Anon. (1916). School Notes. *Magazine of the London (Royal Free Hospital) School of Medicine for Women 11*, 119.
7. Anon. (Jul. 1917). School notes. *Magazine of the London (Royal Free Hospital) School of Medicine for Women, 12*(67), 119.
8. Anon. (1929). *London School of Medicine for Women, Annual Report.*
9. Brown, B. C. (1960). *Isabel fry, 1869–1958: Portrait of a great teacher* (pp. 30–34). A. Barker.
10. Cooper, Y. *Student records*. Archives, Royal Holloway College.
11. Cooper, Y. M. D., & Davies, H. V. (1951). *Aids to practical hygiene for nurses*. Baillière, Tindall & Cox.
12. Williams, M. *Student records*, Archives, Royal Holloway College.
13. Gardner, J. A., & Williams, M. A. (1921). Critical study of the methods of estimating cholesterol and allied substances. *Biochemical Journal, 15*(3), 363–375.
14. Anon. (Mar. 1921). Research work at the LSMW. *Magazine of the London (Royal Free Hospital) School of Medicine for Women, 16*(78), 77–78.
15. Gardner, J. A. & Williams, M. A. (13 Mar. 1922). Quinoline derivatives: Dyes Brit. 198, 462.
16. (1921). Certificates of candidates for election at the ballot to be held at the ordinary scientific meeting. *Proceeding of the Chemical Society, 62*.
17. Anon. (1935). School Notes. *Magazine of the London (Royal Free Hospital) School of Medicine for Women 30*, 63.
18. Anon. (1945/1946). *Royal Free Hospital for Women, University of London, Annual Report.*
19. Anon. (1936). School notes. *Magazine of the London (Royal Free Hospital) School of Medicine for Women, 31*, 76.

20. Anon. (Jul.–Dec. 1947). School notes. *Magazine of the London (Royal Free Hospital) School of Medicine for Women, 9*, 66.
21. Wilson-Smith, M. J. *Student records,* Archives, Royal Holloway College.
22. (1927). Certificates of candidates for election at the ballot to be held at the ordinary scientific meeting. *Proceeding of the Chemical Society, 103.*
23. Anon. (Nov. 1930). College letter, Royal Holloway College Association, 16. RCH AS/902/73.
24. Gee, G. (1952). Ernest Harold Farmer. 1890–1952. *Obituary Notices of Fellows of the Royal Society, 8*, 159–169.

# 16 The End of the LSMW

*In the first part of this chapter, we will look at the challenges which continued for women throughout the first half of the twentieth century when they sought a medical education. However, the major part will be devoted to the environment faced by women students after 1947 when the London School of Medicine for Women was completely absorbed into the co-educational Royal Free Hospital, and then submerged into the Medical School of University College, London. As is vividly highlighted here, the combined institution adopted the misogynistic attitudes of the males. The proud heritage of the London School of Medicine for Women as a haven for women students—and women chemistry instructors—was no more.*

## The First World War

With the commencement of the First World War, enrolment in the male-only London medical Schools plummeted. Admission of women to these Schools came about not through an enlightened attitude towards women, but as a necessity for the survival of many of the men's medical Schools [1]. For example, at Charing Cross Hospital, the historian, R. J. Minney, commented [2]: "They came swarming in. Within a few weeks the male students were mere dots amid the fluttering skirts and flowing hair in the lecture theatre."

However, the welcoming of women students to male-only medical Schools was not to last. Men returned from the War and the next generation of male applicants thronged into the formerly male-only medical Schools. As the 1920s progressed, these medical Schools, one-by-one, reinstated the bar to women. Continuing with the example of Charing Cross Hospital, Minney noted that; with the return of the soldiers at the end of the War, plus new male School graduates [3]: "... not long afterwards women were again barred from the School. It was not until 1948 on the University's insistence, that they were readmitted."



M. Rayner-Canham and G. Rayner-Canham, *Pioneers of the London School of Medicine for Women (1874–1947)*, Perspectives on the History of Chemistry,
https://doi.org/10.1007/978-3-030-95439-0_16

It was the male students at the Medical School of University College, London (UCL), who seemed particularly vociferous in their opposition to the presence of women. Women students had been accepted since 1917; however with other Schools shutting their doors to women, they were spurred into action in October 1920. The men submitted a resolution to the Medical School Committee [4]: "This meeting [of the male students] deplores the action of the Medical School Committee in admitting women students to this Hospital and urges them to refuse them entrance in future."

The accompanying letter expanded upon their views [5]:

> We do ... question the desirability of co-education in our hospital. ... While such hospitals as Guy's and St. Bartholomew's refuse to admit women students we are convinced that they will continue to be a source of attraction to Public School and Oxford and Cambridge men. Again, many second and third year men now in the College are looking with alarm to a time when women will be admitted to House appointments. We need hardly point out, gentlemen, the intolerable position of an ex-Serviceman who has, perhaps, as his House Surgeon over him a girl of twenty-two ...
> We anticipate with the greatest anxiety the future of our hospital, and appeal to you to rescind the decision, taken during the War, before it is too late and the University College Hospital has become the teaching Women's Hospital of the Metropolis.

Carol Dyhouse has convincingly argued that the London male-only hospitals had a masculine culture with an obsession with sports. She commented [6]: "It is difficult to exaggerate the importance of athletics, team sports and particularly rugby in the culture of the medical Schools."

The very presence of women was seen as compromising that male bonding through physical prowess.

## The Inter-War Years

By the end of the 1920s, all the doors to a medical education had re-closed to women, except the London School of Medicine for Women (LSMW) [7] (Fig. 16.1). A commentary in the journal *Nature* stated [8]:

> Prior to the War, [WW1] all the medical schools of the University of London (with the exception of the London School of Medicine for Women) were restricted to men, but it will be remembered that during the War seven of the schools admitted women in addition. These facilities were withdrawn a short time ago, except in the case of University College Hospital, which still admits a limited quota. The action of the authorities of the medical schools aroused considerable discussion, and a Committee was appointed by the Senate of the University of London "to consider the question of the Limitations placed upon the Medical Education of Women Undergraduates." ... The Committee thinks there is no valid argument against co-education, but that co-education to be successful must be voluntary. No countenance is therefore given to the suggestion which has been made that the University should enforce a policy of co-education upon the medical (and other) schools by withdrawal of recognition or other means. Such a policy, to be logical, would have to be applied all round, and this would force men upon women's colleges, and men upon the London School of Medicine for Women!

**Fig. 16.1** Students in the LSMW Organic Chemistry Laboratory, 1924. Public domain: London Metropolitan Archives, unknown photographer

## The End of the LSMW

Thus, the uniqueness of the London School of Medicine for Women (LSMW) remained until the post-Second World War period. It was in 1947 that all institutions were required to become co-educational. The London Royal Free Hospital School of Medicine for Women (LRFHSMW) was required to become co-educational. After being absorbed into the Royal Free Hospital (RFH), the 'for Women' was dropped from its name. It was now the Royal Free Hospital School of Medicine (RFHSM). The *LSMW Magazine* became the *RFH Magazine.*

The importance of team sports for the male students still held true for the first cohorts of arriving male students at the RFHSM. This had not been anticipated as this 1948 Editorial (by a female editor) in the *RFH Magazine* commented [9]:

> When we welcomed our men students into the School last October we did not realise that they might like to play soccer or rugger, or some other exclusively male sport. After some thought, we negotiated with Guy's Hospital Student's Union. The result is that we have now entered into a kind of partnership with Guy's and our men can make use of their male sports clubs, and their women students, our female clubs.

A sign that the values of the LSMW were no more was provided by a 1953 Editorial in the *RFH Magazine.* The view was expressed that the way to raise the status of the School was by increasing the number of male students [10]:

> We are sure that the ladies will forgive the men of this College for maintaining a none too secret notion, that the one sure way to steady progress and increasing eminence of the Medical School (the objective of the progress, and the nature of the eminence is not always clear) is the raising of the number of male students as rapidly as possible. It is perhaps natural for the men to think of the history of this School as beginning in 1947.

Nevertheless, the 1953 Editorial went on to propose that the new institution would be able to reach a true co-educational model before other institutions by virtue of its origin as a women's School [10]:

> For those of us who have entered the School since 1947, it is difficult to imagine the feeling of the School as it then was, for there has been change, ... Sir Henry Dale indicated at the start of the present academic year, that this School might eventually lose its original special character. Few will deny that this is highly probable, but we now have a unique opportunity of eventually attaining a new special character; for we are starting a period of change common to all the Medical Schools from what might be called the "better end" of the scale. From this point we are more likely than are the other schools to reach the happy state in which the number of the sexes are approximately equal, and thus to gain the proud title of "The First Truly Co-Educational Medical School in Britain."

## An Amnesia of History

With the change to a co-educational institution, the contents of the *RFH Magazine* were far different to that of the LSMW Magazine. In particular, a demeaning attitude to women students became prevalent. The first indication was a 'satirical' article titled 'Chemical Analysis of a Woman'. Selected—and very typical—sentences are included below [11]:

> *Physical Properties*
> Surface usually covered with a film of paint or oxide.
> Boils at nothing and freezes without reason.
> *Chemical Properties*
> Highly explosive and dangerous except in experienced hands.
> Possesses affinity for gold, silver, platinum and precious stones.
> Has ability to absorb great quantities of expensive food.
> Sometimes yields to pressure.
> Ages rapidly.
> *Uses*
> Chiefly ornamental.
> Probably most powerful (bank account) reducing agent known.
> It is illegal in most countries to possess more than one specimen.

## Should Women Be Admitted to Medical School?

Yes, this truly was the question of a series of articles in the *RFH Magazine* in 1964. The articles, from a Symposium on Women in Medicine, were reprinted from the *Belfast Hospital Medical Journal* and it was if the preceding 100 years of women's advancement—and the success of the LSMW—had never happened. Here, we will include extracts from each of the articles.

The opening reprinted article by J. H. Biggart (a male doctor) provided an overview of the purpose of the event and whether women should have been admitted to medical Schools [12]:

> Has this been a good thing? That is the question for this symposium. It will be argued that their education is uneconomic, that the loss through marriage is too great, and that in the days of competition for admission it is unfair to keep out male students in their favour. ... All in all it seems wiser to maintain the medical course as an educational one and to continue to allow them to set the standards of hard work and to enliven the lives of their fellow students and their teachers by their presence. ... So bless their little hearts and let them stay.

The second contribution by D. Stephens (a female doctor) initially struck a positive note [13]:

> The question as to whether women should or should not be in Medicine is sheerly hypothetical—they cannot keep us out—we are already in, ... From a purely selfish point of view Medicine is an ideal career for a woman. On one hand if she remains single she has a remunerative and wonderfully satisfying occupation; on the other hand she has a career which is quite compatible with married life. She can bring up her children perhaps doing a Clinic once a week to keep in touch; then when they are at school she can do the occasional Locum, perhaps even join a Group practice. ...

However, after such pro-women sentiment, came an unexpected 'twist' at the end [13]:

> In conclusion, does the community as a whole benefit by the presence of the woman doctor. If there were plenty of doctors the answer would be "Yes". But there are not plenty of doctors. I feel that there should be no increase in the proportion of women to men in Medical School, in fact there should be a decrease. The Medical Schools should be expanded, but the number of women should be strictly limited, and whenever a choice has to be made between a man and a woman of equal ability, the man should get the place every time.

The third contribution to the symposium, also by a female doctor (J. G. Neill), focussed upon the topic of married women doctors. It began by pointing out the more general objection to working women [14]: "The whole question of married women working in any sphere produces emotional reactions which imply that they are taking men's jobs, neglecting their homes and leaving gangs of potentially delinquent children roaming the streets with latchkeys around their necks."

The majority of the article described the author's own experiences and then ended with advice [14]:

> And, finally, perhaps the most important thing is to marry the right husband. It doesn't matter so much what his job is, but he must actively encourage you in your pursuit of work; he must bear cheerfully the crises when domestic help evaporates and the house is further than usual from the Homes and Gardens category, and he must share with you the conviction that the married woman doctor can contribute service of value to the art of medicine.

The fourth article, this one by another male doctor (A. H. G. Love), illustrated how strong the backlash had become to women doctors, *and the reader should note, published in a magazine which had once been for a solely-women School of medicine!* The article began [15]:

> Are women and medicine compatible? They certainly have at least two things in common, they are both complex subjects and they require a lifetime of study. ... I think that it would be best if the modern followers of the Suffragettes did not attempt to use the field of medicine as a battleground on which to establish equality of the sexes.

Dr. Love then pointed out some of the 'shortcomings' of women as doctors [15]:

> Some of the lapses which show women [doctors] in a bad light are I am sure, due to flight of ideas. I think you know what I mean—the sudden awareness of the need to decide what to wear for that date this evening, or the jolted remembrance of an evening meal to prepare without as yet any raw materials. This sort of mental activity is not designed to produce the best in medicine. Women are characteristically emotional creatures and certainly in clinical medicine this does not stand them in good stead.

He concluded with his personal survey of his colleagues as to their attitude to women doctors [15]:

> One last observation is of interest. I put the question—should a woman practice medicine—to some of my staff at coffee the other morning. They were unanimous in condemnation of women in medicine, some even verging on rage phenomena. I wondered why? Was it a memory of a personal dislike or was it a feeling of intrusion in male exclusivism? Whatever it was it seemed a hostile world in which the woman would have to practice her healing art.

The fifth article in the series, by Dr. Bartley, a female doctor, was devoted to a discussion of the specializations adopted by women [16]: "Whatever speciality a woman adopts, it is the considered opinion among medical women that unless her qualifications, achievements and experience are much better than those of any of her male rivals, she is unlikely to be preferred for any post that she is seeking."

## And a Final Disappearance

Then in 1996, the Royal Free Hospital School of Medicine was, itself, merged into the Medical School of University College, London. Finally, in 2008, after another merger, its name disappeared to become simply the UCL Medical School. Ironic, as it had been the UCL students of 1920 who had been so vociferous in their objection to the admission of women [4]. The pioneering LSMW, their generations of women students, and the enthusiastic women chemists who taught there, were now buried beneath this series of mergers, and long forgotten.

## References

1. Dyhouse, C. (1998). Women students and the London medical schools, 1914–1939: The anatomy of a masculine culture. *Gender and History, 10*(1), 110–132.
2. Minney, R. J. (1967). *The two pillars of Charing Cross: The story of a famous hospital* (p. 153). Cassell.
3. Ref. 2, Minney, p. 154.
4. Merrington, W. R. (1976). *University College Hospital and its medical school: A history* (p. 237). Heinemann.
5. Ref. 4, Merrington, pp. 238–239.
6. Ref. 1, Dyhouse, p. 125.
7. Dyhouse, C. (1998). Driving ambitions: Women in pursuit of a medical education, 1890–1939. *Women's History Review, 7*(3), 321–343.
8. Anon. (9 Feb. 1929). News and views. *Nature, 123*, 217.
9. Anon. (Jan.-Jun 1948). Editorial. *The Magazine of the Royal Free Hospital School of Medicine, 10*(23), 1.
10. Anon. (1953). Editorial. *The Royal Free Hospital Magazine, 39*, 45.
11. Anon. (Dec. 1955). Chemical analysis of a woman. *The Royal Free Hospital Magazine, 17*(43), 105.
12. Biggart, J. H. (Mar. 1964). Women in medicine. *The Royal Free Hospital Magazine, 26*(81), 138–139.
13. Stephens, D. (Mar. 1964). Battle of the sexes. *The Royal Free Hospital Magazine, 26*(81), 139–140.
14. Neill, J. G. (Mar. 1964). Married women doctors. *The Royal Free Hospital Magazine, 26*(81), 140–141.
15. Love, A. H. G. (Mar. 1964). Should women practice medicine? *The Royal Free Hospital Magazine, 26*(81), 142–143.
16. Bartley, Dr. (Mar. 1964). Women and specialization in medicine. *The Royal Free Hospital Magazine, 26*(81), 143–145.

# Index






M. Rayner-Canham and G. Rayner-Canham, *Pioneers of the London School of Medicine for Women (1874–1947)*, Perspectives on the History of Chemistry,
https://doi.org/10.1007/978-3-030-95439-0

Printed in Great Britain
by Amazon